STUDENT UNIT GUIDE

NEW EDITION

OCR A2 Biology Unit F214
Communication, Homeostasis and Energy

Richard Fosbery

PHILIP ALLAN

Philip Allan, an imprint of Hodder Education, an Hachette UK company, Market Place, Deddington, Oxfordshire OX15 0SE

Orders
Bookpoint Ltd, 130 Milton Park, Abingdon, Oxfordshire OX14 4SB
tel: 01235 827827
fax: 01235 400401
e-mail: education@bookpoint.co.uk
Lines are open 9.00 a.m.–5.00 p.m., Monday to Saturday, with a 24-hour message answering service. You can also order through the Philip Allan Updates website: www.philipallan.co.uk

© Richard Fosbery 2012

ISBN 978-1-4441-6254-7

First printed 2012
Impression number 5 4 3
Year 2017 2016 2015 2014

Cover photo: Fotolia
Other photos © Richard Fosbery

Printed in Dubai

Hachette UK's policy is to use papers that are natural, renewable and recyclable products and made from wood grown in sustainable forests. The logging and manufacturing processes are expected to conform to the environmental regulations of the country of origin.

Contents

Getting the most from this book

Questions & Answers

Exam-style questions

Examiner comments on the questions

Tips on what you need to do to gain full marks, indicated by the icon ⓔ.

Sample student answers

Practise the questions, then look at the student answers that follow each set of questions.

Examiner commentary on sample student answers

Find out how many marks each answer would be awarded in the exam and then read the examiner comments (preceded by the icon ⓔ) following each student answer.

About this book

This unit guide is the first of two that cover the OCR A2 Specification in biology. It is intended to help you prepare for **Unit F214: Communication, Homeostasis and Energy**. It is divided into two sections:

- **Content Guidance** — here you will find key facts, key concepts and links with other parts of the AS/A2 biology course. You should find the **Focus on practical skills** sections useful for the practical work assessed in Unit F216. There are sections on histology to help you interpret photographs, drawings and diagrams of sections of the pancreas, liver and kidney. The **synoptic links** are intended to show you how topics in this unit build on information you learnt at AS. It also shows you how information in this unit links to topics in Unit F215: Control, Genomes and Environment.
- **Questions and Answers** — here there are questions on each of the six sections in Unit F214, together with answers written by two candidates and examiner's comments.

This is not just a revision aid. This is a guide to the whole unit and you can use it throughout the A2 course. You should read other sources of information such as textbooks, articles from *Biological Sciences Review*, published by Philip Allan Updates, and websites. It is also a good idea to use animations available on the internet to follow the complex processes described in this unit guide. I have recommended some although I am sure you will be able to find others.

In this guide there are references to the three tasks you will take as part of Unit F216:

- the **qualitative task**, e.g. carrying out an experiment that does not give you anything to measure or determine — it may involve recording colours or drawing from a specimen or from a microscope slide
- the **quantitative task**, e.g. carrying out a practical task in which you record measurements
- the **evaluative task**, e.g. commenting critically on the practical procedure and the results you obtained in the quantitative task

There is a Student Unit Guide specifically for the practical assessment in F213. You will find some references to it in the sections entitled 'Focus on practical skills' in this guide for F214.

Units

Various units are used in this guide.

Length

The units used are nm, μm, mm, m and km:

- 1000 nm (nanometres) = 1 μm (micrometre)
- 1000 μm = 1 mm (millimetre)
- 1000 mm = 1 m
- 1000 m = 1 km

In this unit you will study both cells (for example nerve cells and endocrine cells) and organelles (for example chloroplasts and mitochondria). In the unit test you may be asked to apply your knowledge of calculating magnification and size from Unit F211 to drawings, photomicrographs or electron micrographs of cells and/or organelles. Wavelengths of light are measured in nanometres. You will need to use these units in the section on photosynthesis.

Volume

The units used are cm^3 and dm^3: $1000\,cm^3 = 1\,dm^3$.

You will often find ml (millilitre) on glassware and in textbooks. Examination papers, however, use cm^3 (cubic centimetre or 'centimetre cubed') and dm^3 (cubic decimetre or 'decimetre cubed'). $1\,cm^3$ is the same as $1\,ml$; $1\,dm^3$ is the same as 1 litre ($1\,l$ or $1\,L$); $1000\,cm^3 = 1\,dm^3$.

In this guide you will come across volumes in the sections on excretion, photosynthesis and respiration.

Energy

The units used are joules (J), kilojoules (kJ) and megajoules (MJ):
- $1000\,J = 1\,kJ$
- $1000\,kJ = 1\,MJ$

On p. 70 you will find the energy values of some respiratory substrates. You may be asked to recall these figures in the unit test and give the appropriate unit (the kilojoule).

Potential difference

The units used are volts (V), millivolts (mV): $1000\,mV = 1\,V$.

Millivolts are used in the section on nerves (pp. 15–27). There are two figures to remember: the potential difference across a membrane at rest is between $-65\,mV$ and $-70\,mV$. This is known as the resting potential. During an action potential, the potential difference increases to between $+30\,mV$ and $+40\,mV$.

Time

The units used are hours (h), minutes (min), seconds (s) and milliseconds (ms): $1000\,ms = 1\,s$.

Milliseconds are used to measure the speed of transmission of nerve impulses. In recording results from investigations into the rates of photosynthesis and respiration you will probably measure time in seconds. You can calculate rates by using the formula $1/t$ where t = the length of time it takes to collect a certain volume of gas or for a colour change to take place.

Names of biochemical molecules

Much of this unit deals with biochemistry. Almost all biochemical molecules have more than one name. The names used in this guide follow those used in the specification. Organic acids, such as pyruvic acid, acetic acid and citric acid exist as anions in cytoplasm and body fluids and are referred to using the suffix '-ate' — hence pyruvate, acetate and citrate. In several places the names acetic, acetate and acetyl are used, which you may also know as ethanoic, ethanoate and ethanoyl. Adrenaline and noradrenaline are now known officially as epinephrine and norepinephrine. You should use these words if you do a web search for extra information or for help with your revision. However, the specification refers to adrenaline and so this term and noradrenaline are used in this guide.

Phosphate ions (PO_4^{3-}) are involved in many biochemical processes. In this guide you will read about them in the sections on excretion, photosynthesis and respiration. P_i is a standard abbreviation for inorganic phosphate and is used in this guide.

A2 biology

The diagram below shows you the three units that make up the A2 course. You should have a copy of the specification for the whole of the course. You should keep it in your file with your notes and refer to it constantly. You should know exactly which topics you have covered so far and how many more you have to do.

Unit F214		Unit F215		Unit F216
Communication, Homeostatis and Energy	+	Control, Genomes and Environment	+	Practical Skills in Biology 2

The specification outlines what you are expected to learn and do. The content of the specification is written as **learning outcomes**, which state what you should be able to do after studying and revising each topic. Some learning outcomes are very precise and cover just a small amount of factual information. Some are much broader. Do not think that any two learning outcomes will take exactly the same length of time to cover in class or during revision. Some of the learning outcomes deal with practical biology — in this guide these are covered in sections called 'Focus on practical skills'.

It would be a good idea to write a glossary to the words in the learning outcomes, as the examiners expect you to know what they mean. This guide should help you to do this. You have to use your AS knowledge during the course, so links to AS are given throughout the guide.

Themes in biology

The first theme in F214 is homeostasis, which is dealt with in the context of thermoregulation. Later, you meet homeostasis in the control of glucose concentration and water content in the blood. The second theme is communication by the nervous and hormonal systems. The third is energy in living organisms, covered by photosynthesis and respiration. Although the specification refers almost exclusively to mammals and flowering plants, remember that many of the processes described in this guide occur in other organisms — for example, homeostasis occurs within yeast cells as well as in humans.

Examiners have to set some challenging questions in the unit test. Identifying the main themes in this unit and making links between different topics are two ways in which you can prepare for these questions.

This guide will help you to understand how the key facts fit into the wider picture of biology — you need knowledge of this unit to understand fully some of the content of Unit F215. You also need to revise topics from AS — they provide useful background material because many of the topics at A2 build on the subject matter covered at AS. Some links to AS are made clear at the beginning of each section and others are explained in the **synoptic links** at the end of each section. In the unit test you will be assessed on your ability to apply knowledge and understanding of more than one area of the specification to a particular situation or context. This means that you should revise the material you studied at AS and be prepared to use it when learning new topics and answering questions. This guide will help you do this by identifying these synoptic links.

Content Guidance

Module 1: Communication and Homeostasis

Communication

Key concepts you must understand

If they are to survive, all organisms must detect **stimuli** (changes) in their surroundings and respond to them. Complex, multicellular organisms are composed of specialised cells, tissues, organs and organ systems. Cells communicate with each other by secreting chemicals that diffuse into the immediate surroundings and are detected by adjacent cells. However, this method is only effective over short distances. So that cells throughout the body can function efficiently and effectively, organisms need long-distance communication systems to coordinate and synchronise their activities. Animals have two such systems:

- the nervous system
- the endocrine (hormonal) system

Four ways in which cells signal to one another are:

(a) paracrine secretion, over short distances
(b) nerve cells use electrical impulses to cover long distances, but release chemicals where they terminate
(c) endocrine secretion
(d) neurosecretion

In endocrine secretion and neurosecretion, chemicals are released into the blood to travel long distances.

To function efficiently organisms have control systems to keep internal conditions near constant, a feature known as **homeostasis**. This requires information about conditions inside the body and in the surroundings. These conditions are detected by sensory cells. Some of the physiological factors controlled in homeostasis in mammals are:

- core body temperature
- blood glucose concentration
- concentration of ions in the blood, e.g. Na^+, K^+ and Ca^{2+}
- water potential of the blood

Examiner tip
These four methods of signalling are shown in Figure 1. As you look at the diagram remember what you learnt about cell signalling and cell surface receptors in Unit F211.

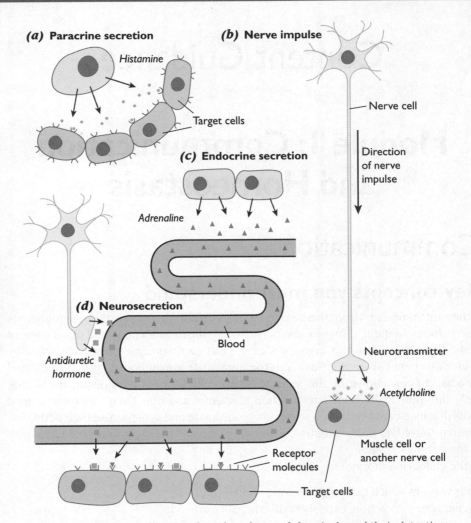

Figure 1 Cell signalling involves the release of chemicals and their detection

Key facts you must know

Homeostasis and thermoregulation

Homeostasis is the maintenance of near constant internal conditions. It involves detecting changes inside and outside the body and making responses that counteract changes in the internal environment. Birds and mammals maintain a near constant core body temperature independent of the temperature of their surroundings. They are **endotherms** — they generate heat and conserve it in their bodies if it is cold, and have mechanisms to lose heat if they are hot. Other animals and plants are unable to do this and are known as **ectotherms**. Their main source of body heat is from their surroundings, and their body temperatures fluctuate with the ambient temperature. It is a mistake to call ectotherms, such as lizards, snakes and fish 'cold-blooded'. On a hot day a lizard is hot-blooded and tropical fish are definitely warm-blooded. Although ectotherms are dependent on the temperature of their surroundings, many organisms are able to maintain a temperature different from that of their surroundings.

- Some lizards are able to regulate their temperature using behavioural methods such as basking in the sun and moving into the shade.
- Desert plants are often a few degrees cooler than the air temperature on hot days.
- Arctic and Antarctic plants and insects are often dark to absorb heat so that their temperatures are several degrees higher than air temperature.

The normal range of human body temperature is 36.0–37.6°C. Most birds have higher temperatures, for example 40–42°C.

In order to regulate temperature the body needs:
- receptors that detect changes in the temperature of the surroundings and in the internal (core) body temperature
- a control centre that receives information about changes in temperature and sends instructions to the body to make adjustments to counteract changes in body temperature
- effectors that produce heat, promote heat loss or promote heat conservation

The thermoregulatory centre is situated in the **hypothalamus** in the brain. It lies just above the pituitary gland in the centre of the head. The anterior (front) part of the hypothalamus responds to increases in temperature by promoting heat loss. The posterior (back) part responds to decreases in temperature by conserving heat and stimulating heat production.

The hypothalamus constantly compares information about the temperature of the surroundings and the internal temperature with the set point and acts to keep the difference between the *actual* body temperature and the set-point temperature as small as possible. The control centre needs information about whether instructions sent to effectors to carry out corrective actions to produce heat, lose heat or gain heat are effective. This involves feedback in which a change in the factor being controlled acts as a stimulus that leads to a response which counteracts the change in the factor. The effects of corrective actions are monitored by the hypothalamus and adjustments are made to keep the temperature within narrow limits.

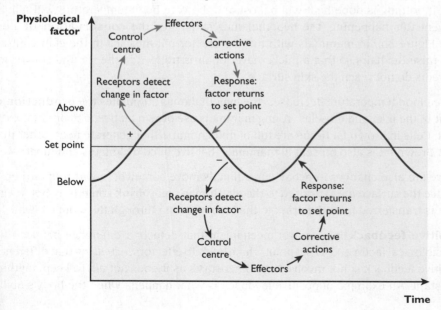

> **Knowledge check 2**
>
> Define the term *homeostasis*.

> The **set point** is the value of the physiological factor that the body tries to keep constant all the time.

Figure 2 The principle of negative feedback

> **Knowledge check 3**
>
> Name the thermoregulatory centre in the brain.

This control method is called **negative feedback** (see Figure 2). Negative feedback mechanisms keep a physiological factor, such as core body temperature, within narrow limits. The set point for temperature can change according to conditions (see the first part of Question 1 on p. 73).

The flow charts In Figure 3 show how the hypothalamus is involved in controlling body temperature in humans. Part (a) of the flow chart shows what happens when you absorb heat or generate heat internally during exercise. Part (b) shows what happens when the ambient temperature is low and you lose heat to your surroundings. Apply the key terms to each part of the flow chart by writing an explanation of the control of body temperature.

When the temperature of the surroundings increases, the skin absorbs heat. This stimulates peripheral (outer as opposed to central) temperature receptors in the skin. During exercise, heat is released from muscles and increases the blood temperature. This increase is detected by central temperature receptors in the spinal cord and in the hypothalamus. When the body temperature rises for either or both of these reasons, the hypothalamus stimulates the **loss of heat** (see Figure 3a). Many mammals have few sweat glands and are covered in fur. In addition to the changes shown, in many mammals the heat loss centre also stimulates:

- erector pili muscles in the skin to relax so that hairs lie flat and less air is trapped by the fur
- shallow breathing (panting), which causes loss of heat by evaporation of water from the mouth and gas exchange system

There are also changes in behaviour: animals move into the shade or into a burrow, they are inactive so they do not produce heat through muscle contraction, they avoid the hottest part of the day and they may drink something cold so that heat is transferred from the blood to water in the stomach.

When the temperature of the surroundings decreases, temperature receptors in the skin are stimulated. They send information to the hypothalamus — early warning that if nothing is done heat will be lost and the core body temperature will fall. To prevent this happening, the hypothalamus stimulates the **conservation of heat** (see Figure 3b). In mammals with fur, the erector pili muscles in the skin contract and raise the hairs so that a thicker layer of air is trapped in the fur and convection currents do not reach the skin surface.

If the blood temperature decreases, the hypothalamus stimulates the **production of heat** in the liver and muscles. Young mammals use brown fat tissue to produce extra heat. Cells in brown fat tissue are full of mitochondria that generate heat rather than ATP. Brown fat is also present in mammals that live in cold climates and hibernate.

There are also changes in behaviour. Animals move somewhere warm or curl up to reduce the surface area exposed to the air. Humans may drink something hot so that heat is transferred from the drink to the blood flowing through the stomach wall.

Positive feedback is a control mechanism that detects a difference between the physiological factor and its set point which leads to effectors *increasing* that difference. Positive feedback is not involved in homeostasis as it does not act to keep anything constant. An example of positive feedback is what happens when the body's ability

Examiner tip

When you write answers about homeostasis you should try to use the following key terms: negative feedback, set point, control centre, receptor, effector and corrective actions.

Examiner tip

You learnt about the latent heat of vaporisation of water in Unit F212. As you read through this guide you should look for links with concepts you studied in the AS units. Get used to looking up these concepts and adding such information to your notes.

Knowledge check 4

Explain briefly how an endothermic animal, such as a bird or mammal, prevents its core body temperature decreasing to the ambient temperature on a cold day.

Examiner tip

The key message here is that if you get cold, the corrective actions you take attempt to conserve the heat present in your body before using energy reserves to generate more heat.

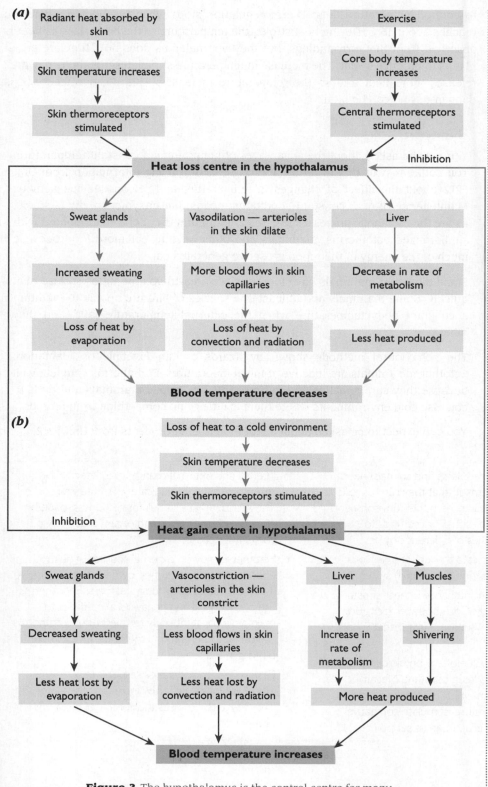

Figure 3 The hypothalamus is the control centre for many aspects of homeostasis, including body temperature

Examiner tip

Positive feedbacks are not sustainable. They happen for a short time and then stop, as in action potentials in nerve cells (see p. 20).

Knowledge check 5

Explain the difference between control by negative feedback and control by positive feedback.

to control temperature fails. Thermoregulation stops when the body temperature reaches about 45°C. During heat stroke, the temperature of the body rises as heat is absorbed from the surroundings, but the hypothalamus does not stimulate sweat glands to produce sweat. The increase in temperature stimulates metabolism, which releases more heat and so the temperature increases. This positive feedback is dangerous and can be fatal.

Synoptic links

You may be asked about the advantages of maintaining a constant temperature. You could answer this in terms of enzyme action using information from Unit F212 about the effect of changes in temperature and pH on enzyme activity. Maintenance of near constant conditions means that enzyme activity is stable, as are the metabolic processes (e.g. respiration) that they control. Compared with similar-sized ectotherms, mammals and birds have to eat more food because much of the energy in their food is used to generate heat.

Small birds and mammals have large surface area-to-volume ratios so lose heat quickly. Some mammals and at least one species of bird are unable to maintain a constant body temperature when the external temperature falls, and they hibernate.

The behavioural methods shown by lizards are an example of adaptation. Ectothermic animals are inactive at low temperatures. Endotherms compete well because they can be active all the time. This has helped mammals and birds to colonise cold environments where there is little or no competition from reptiles.

You can expect to be asked about adaptation to environments from Unit F212.

Summary

- Multicellular organisms (plants and animals) need communication systems so that they can respond to internal changes and changes in their environment. These systems use cell signalling methods to coordinate the activities of different organs.

- Animal cells communicate through the nervous system and the endocrine (hormonal) system.

- Homeostasis is the maintenance of near constant conditions in the body. Core body temperature is an example of a physiological factor that is kept within narrow limits

- The set point is the value of any physiological factor that the body tries to keep constant. Negative feedback is the way in which changes in a factor, such as body temperature, stimulate corrective actions to restore the factor to its set point.

- Internal and external changes are detected by receptors that communicate with a central control. These changes are stimuli for responses made by effector cells — tissues and organs that carry out corrective actions.

- Ectotherms rely on external sources of heat; endotherms generate their own heat and conserve it in their bodies using fur or feathers. Endotherms use physiological and behavioural methods to maintain a constant body temperature; ectotherms use primarily behavioural methods, for example basking in the sun.

- Positive feedback systems respond to a change in a physiological factor by increasing the factor. These systems, such as the control of birth, are not homeostatic.

Nerves

Key concepts you must understand

The nervous system consists of the central nervous system (CNS) and the peripheral nervous system (PNS). The CNS is composed of the brain and spinal cord. The PNS consists of cranial nerves that are attached to the brain, and spinal nerves that are attached to the spinal cord. Within the nervous system there are nerve cells that have different functions. Functionally, the nervous system is divided into:

- the voluntary nervous system
- the involuntary nervous system

Both systems consist of nerves from sensory receptors and nerves to effectors (muscles and glands). The voluntary system sends sensory information about the surroundings to the CNS and controls conscious actions. The involuntary system sends sensory information about the internal environment to the CNS and controls the activity of organs such as the heart and stomach.

Sensory receptors are **transducers** that convert one form of energy into another. Receptors are adapted to detect specific forms of energy. For example, chemoreceptors in the nose and tongue have receptor molecules on their cell surface membranes that have shapes complementary to those of the molecules they detect. Rod and cone cells in the retina in the eye have light-sensitive pigments that absorb light energy.

Nerve cells are specialised cells that transmit electrical impulses over long distances. One end of a nerve cell is specialised to receive information and the other end is specialised to send information. This means that within the peripheral nervous system impulses are sent either from sensory receptors to the CNS or from the CNS to effectors. Impulses travel along neurones in one direction.

Key facts you must know

A nerve is a bundle of nerve cells or **neurones** surrounded by fibrous tissue. A section of a nerve is illustrated in Figure 4.

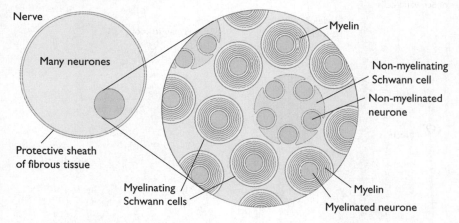

Figure 4 A section of a nerve as seen using a light microscope

Examiner tip

The involuntary nervous system is involved in many aspects of homeostasis, such as temperature control (see pp. 10–14).

Examiner tip

The term *complementary* is often used in biology. Remember what you learnt in F212 about enzymes and substrates. There are many other examples of molecules 'fitting together'. Try listing them, as this will help your synoptic understanding.

Knowledge check 6

Explain why receptors are described as *transducers*.

Examiner tip

Make sure that you can distinguish between a nerve and a nerve cell (neurone). Most of what follows in this book is about neurones. When writing answers, it is better to use the term 'neurone', rather than 'nerve cell' in case you forget to include the word 'cell'.

Examiner tip

You may not have heard of glial cells before, although you should have heard of Schwann cells. Glial cells have important roles other than providing protection and insulation. For example, they are involved in removing neurotransmitters from synapses and maintaining potassium ion concentrations in the tissue fluid around neurones.

Knowledge check 7

The speed of conduction in neurones varies between $1\,\text{m}\,\text{s}^{-1}$ and $120\,\text{m}\,\text{s}^{-1}$. Explain briefly the factors that influence conduction speeds in neurones.

Neurones in nerves and in the spinal cord and brain are supported by **glial cells**. All the neurones in peripheral nerves are surrounded by special glial cells known as Schwann cells. Myelinated neurones are insulated by myelin made by Schwann cells and they transmit impulses very quickly. Non-myelinated neurones transmit impulses at much slower speeds. **Sensory neurones** (see Figure 5a) transmit impulses from sensory cells to the CNS. **Motor neurones** (see Figure 5b) transmit impulses from the CNS to muscles and glands. The cell bodies of sensory neurones are in swellings (ganglia) situated just outside the spinal cord. The cell bodies of many motor neurones are inside the spinal cord. Some motor neurones in the involuntary nervous system have their cell bodies in ganglia outside the spinal cord, for example in the solar plexus.

Myelin is composed of concentric layers of cell membrane arranged like a Swiss roll. As they develop, Schwann cells wrap themselves around the neurone to give layers of cellular material (see Figure 5c). Each layer of myelin consists of two membranes with a very thin layer of cytoplasm sandwiched between them. The membranes are composed mostly of phospholipids; there are very few proteins. The inner layer of myelin is so close to the neurone that tissue fluid and the ions it contains do not reach its cell surface membrane. Tissue fluid only makes contact with the cell surface membrane at the nodes of Ranvier, which are gaps between the Schwann cells.

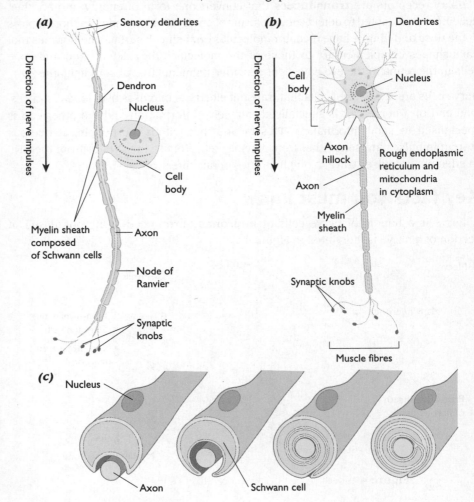

Figure 5
(a) A sensory neurone
(b) A motor neurone
(c) A Schwann cell making myelin by wrapping around a neurone

Table 1 Myelinated and non-myelinated neurones

Feature	Myelinated neurones	Non-myelinated neurones
Schwann cells surrounding neurones	Produce myelin, concentric layers of membrane rich in lipid that act as insulation, preventing ions from reaching the surface membrane of the neurone	No myelin produced
Action potentials	Occur only at nodes of Ranvier	Occur along the whole length of the neurone
Conduction of impulse	Saltatory ('jumping')	Contiguous — wave-like along the whole neurone
Typical speed of impulse transmission/m s^{-1}	100–120	10–20
Distribution in animal groups	Vertebrates	Invertebrates (e.g. squid and earthworm); vertebrates
Roles in mammals	Impulse transmission in the voluntary nervous system, e.g. controlling contraction of skeletal muscles in walking (see Figure 6); rapid transmission in the CNS, e.g. up and down spinal cord	Impulse transmission in the involuntary system, e.g. controlling the heart rate (see Figure 18 on p. 36)

Neurones do not bring about behaviour on their own — they work in circuits. The simplest piece of behaviour we can show is a reflex action. Figure 6(b) shows the knee-jerk reflex, which is a monosynaptic reflex. Stretch reflexes like the knee-jerk reflex are involved in maintaining posture and balance and are used during walking. The diagram shows that when the muscle spindle is stimulated by stretching the tendon, impulses travel to the spinal cord. The impulse is then sent across a synapse to the motor neurone that sends impulses to the quadriceps femoris muscle in the leg, which contracts. These two neurones work in series (one after the other). The circuit is known as a **reflex arc**. In many other reflex arcs there is a relay (connector) neurone between the sensory and motor neurones. Relay neurones are found only within the grey matter of the spinal cord and the brain. They allow connections within the nervous system that may influence the reflex action, for example inhibit it (see p. 26).

Transmission of impulses

There are some animations of the nerve impulse on YouTube, but try the animations at http://highered.mcgraw-hill.com/sites/0072943696/student_view0/chapter8.

An impulse is the flow of current along a neurone. This is similar to current flow in a wire, except that current is carried longitudinally along a neurone by the flow of positive ions, not by the flow of electrons. Current decays because of the resistance provided by cytoplasm and by the leakage of positive ions out through the membrane into tissue fluid. In very short neurones, such as those in the brain and in the retina of the eye, flow of current is sufficient to carry information along the length of the neurone. However, if the distance is more than 1–2 millimetres the current decays and has to be 'boosted' at intervals along the neurone. This 'boosting' effect is an **action potential** and is how an impulse is propagated along a neurone. In non-myelinated neurones boosting occurs all the way along the neurone to give contiguous flow.

Examiner tip

Contiguous means close together. Action potentials in non-myelinated neurones occur all along the neurone. In myelinated neurones, action potentials occur only at nodes.

Knowledge check 8

Distinguish between the roles of sensory, motor and relay neurones.

Examiner tip

It is well worth finding some good animations of the conduction of nerve impulses so you appreciate the importance of action potentials and how they are propagated along neurones. Take notes as you watch them. For an impulse to travel the length of a neurone, action potentials must occur at intervals. The word propagation means 'produce more' and is used to mean the production of action potentials that are repeated all along the neurone.

In myelinated neurones, the 'booster' effect occurs only at the nodes of Ranvier where the axon membrane is exposed to the tissue fluid. Impulse transmission in myelinated neurones is saltatory ('jumping') and is faster than contiguous transmission in non-myelinated neurones.

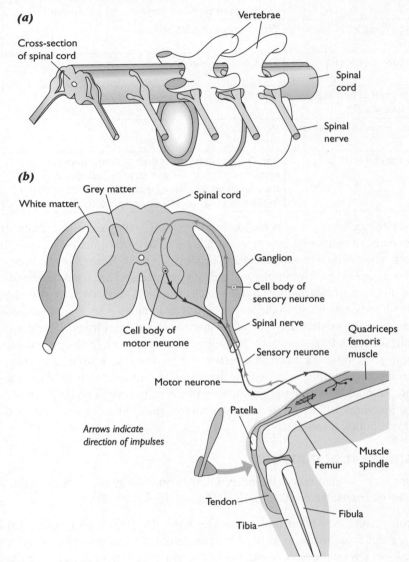

Figure 6 (a) The position of the spinal cord inside the vertebral column and the origin of spinal nerves (b) The monosynaptic reflex arc that coordinates the knee-jerk reflex

Figure 7 shows the important components of the membrane of an axon. Leak channel proteins are open all the time and are selective for either sodium or potassium ions. There are many more of these channels for potassium ions than for sodium ions. Voltage-gated ion channels are shut when the axon is not conducting an impulse. When stimulated, their activation gates open to let ions flow through the membrane. The gates on the sodium ion channels open faster than those on the potassium ion channels. Voltage-gated sodium ion channel proteins have an additional inactivation gate that closes and stays shut for a while before opening again. Pump proteins use ATP to pump sodium ions out and potassium ions in to the axon.

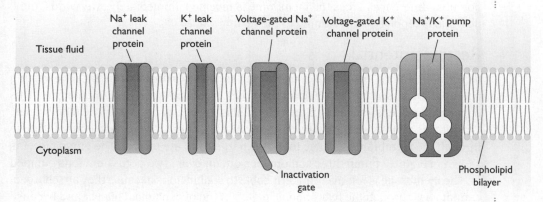

Figure 7 The neuronal membrane contains leak and channel proteins for facilitated diffusion, and sodium–potassium pump proteins for active transport. Both types of voltage-gated channel proteins have activation gates; only the sodium channel protein has an inactivation gate.

Respiration occurs continuously to provide energy to pump ions and maintain concentration gradients. This is like a constantly recharging battery.

The effect of these pumps and channels can be seen when an impulse passes along a neurone. The electrical activity in neurones is detected by placing electrodes at intervals across a neuronal membrane (see Figure 8).

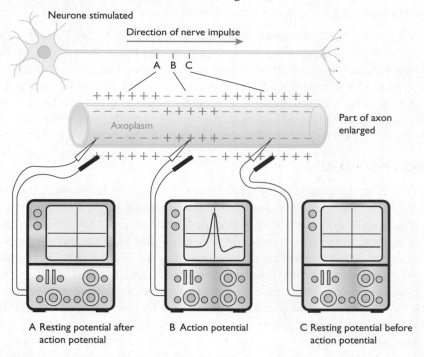

Figure 8 Changes in potential difference across neuronal membranes are detected using microelectrodes and displayed on oscilloscopes

Examiner tip

During an action potential, charge is reversed across the neuronal membrane. This is because sodium ions move into the neurone from the surrounding tissue fluid. The effect of this is to 'boost' current flow along the neurone. Action potentials are often called 'spikes' because of the way they appear on oscilloscopes.

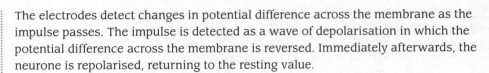

The electrodes detect changes in potential difference across the membrane as the impulse passes. The impulse is detected as a wave of depolarisation in which the potential difference across the membrane is reversed. Immediately afterwards, the neurone is repolarised, returning to the resting value.

Resting potential

When not sending an impulse, a neurone is at a **resting potential**. The resting potential is caused by the unequal distribution of ions across the membrane. This happens in all cells, including plant cells, and is a property of cell surface membranes. Cytoplasm contains organic anions (negatively charged ions) that cannot diffuse through the membrane because they are charged and are too big. The pump proteins shown in Figure 7 pump three sodium ions out for every two potassium ions pumped in. The membrane is relatively impermeable to sodium ions because they are charged, cannot pass through the hydrophobic region of the phospholipid bilayer and because there are few leak channel proteins specific for these ions. Voltage-gated channel proteins for sodium ions and potassium ions are shut when neurones are at rest.

The potential difference across the membrane is between –60 mV and –70 mV.

This means that the *inside* of the membrane is negatively charged with respect to the *outside*. The actual resting potential depends on the concentration of potassium ions in the tissue fluid around the neurone. There are more leak channel proteins for potassium ions than there are for sodium ions and as they are open all the time they allow potassium ions to flow out down their concentration gradient. However, the internal negative charge attracts many potassium ions so they tend to remain inside the neurone giving a high intracellular concentration.

At this stage, the neurone can be compared with a battery before current starts to flow in the circuit. The neurone has a store of energy in the form of concentration gradients for ions. In addition, the inside of the neurone has a negative charge so there is an electrical gradient attracting positive sodium ions into the neurone. The sum of these two gradients is called the **electrochemical gradient**. There is also a concentration gradient for potassium ions as there is a higher concentration of these inside the neurone. These gradients are put to work during the passage of an impulse.

Action potential

There are special channels in the axon hillock that open in response to the stimuli received from other neurones. If the stimulus is great enough, the axon hillock is depolarised sufficiently so that an **action potential** occurs. This generates current flow along the neurone in the direction of the effector. This gives a higher concentration of positive ions in the region immediately adjacent to the axon hillock. The membrane is depolarised as the potential difference is reduced from –70 mV to –50 mV. Voltage-gated sodium ion channel proteins are sensitive to this depolarisation. The arrival of positively charged ions is enough to trigger the activation gates of some of these channel proteins to open allowing sodium ions to flow across the membrane *into* the neurone. As this happens, the potential difference becomes less negative. This triggers more of these sodium channels to open. The potential difference continues to change, triggering yet more to open. This is a **positive feedback** in that a disturbance stimulates more and more channels to open causing an explosive change

OCR A2 Biology

in the potential difference. This only happens for a very short time and not many ions flow in. Voltage-gated sodium ion channel proteins remain open for a short while and then shut firmly as their inactivation gates close. The potential difference reaches +30 mV to +40 mV (see Figure 9a).

To return the potential difference to −70 mV, voltage-gated potassium ion channel proteins respond to the change in potential difference and open, allowing potassium ions to flow out of the neurone down their concentration gradient. This restores the resting potential very quickly.

Figure 9 shows that the high conductance for sodium ions coincides with the rising phase of the action potential. This is when the voltage-gated sodium ion channel proteins are open and sodium ions diffuse into the axon down an electrochemical gradient. The falling phase coincides with the inactivation of these channel proteins when they shut. It also coincides with the high conductance for potassium ions when the voltage-gated potassium ion channel proteins open, allowing potassium ions to diffuse out and the membrane to repolarise. There is an 'undershoot' to −90 mV because these channel proteins are slower to close than the sodium channel proteins; more potassium ions diffuse out than necessary to return the potential difference to −70 mV.

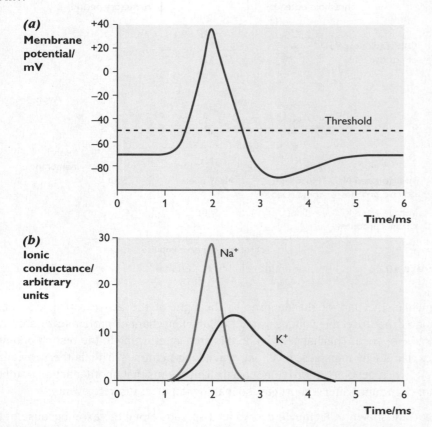

Figure 9 The change in (a) potential difference and (b) the conductance of sodium and potassium ions during the passage of an action potential

Propagating the action potential

At the height of the action potential there is a greater concentration of positively charged ions inside the neurone. These boost the current flow to the next part of the neurone. Current flow is illustrated in Figure 10. The impulse is shown travelling from right to left. The active region is the part of the axon at the height of the action potential with a membrane potential of +40 mV. The inflow of sodium ions increases the concentration of positively charged ions (mainly K^+). These repel each other and are attracted by large negatively charged anions so they spread outwards to left and right inside the axon. On the outside of the axon there is a slight negative charge as sodium ions have moved into the axon. This attracts positively charged ions (mainly Na^+) from both sides that flow in the tissue fluid. This flow of current (as positively charged ions on both sides of the membrane) depolarises the next patch of membrane to the left. This depolarisation is the beginning of the rising phase of the action potential and always reaches the threshold value, stimulating the opening of many voltage-gated sodium ion channels.

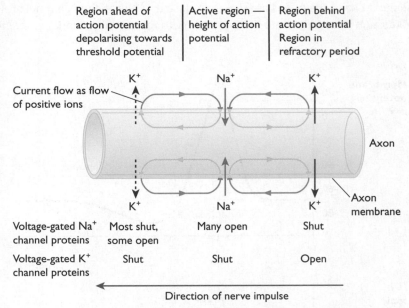

Figure 10 Current flow in a neurone in front of and behind the region with an action potential

Meanwhile, the part of the neurone to the right of the active region cannot be stimulated because the voltage-gated sodium ion channel proteins are shut and cannot be opened. This happens while this area is returning to the resting potential and explains why impulses travel one way along neurones. This is the refractory period, when depolarisation cannot occur. This means that another action potential cannot occur until after a short gap, so making impulses discrete events.

The current shown in Figure 10 decays as it spreads along the axon because of the resistance of the cytoplasm and because some potassium ions diffuse out through leak channel proteins that are always open. Remember that the purpose of the action potential is to boost this current flow so an impulse can travel along a neurone without decaying.

Examiner tip

Why always? Remember that once an action potential starts at the axon hillock it is repeated all the way down the neurone — the **all-or-nothing** response (see p. 24).

Examiner tip

Remember that Na^+ is the main extracellular cation (in tissue fluid) and K^+ is the main intracellular cation.

Refractory means stubborn or unresponsive. The refractory period is the key to understanding how action potentials are discrete events that do not merge into one another. Drum your finger on a table at a regular frequency. Change the frequency. This is a model for information encoding by neurones.

Knowledge check 10

Explain why a nerve impulse travels in one direction along a neurone.

Action potentials do not vary in size — they all show the same amplitude (or 'spike'). However, the **frequency** at which they are sent along neurones varies. If the stimulus is large, there is a high frequency of impulses. A lower stimulus leads to a lower frequency of impulses. So at **A** on Figure 8 the oscilloscope would record a large number of impulses per second with a large stimulus, but a small number per second with a low stimulus (see Question 2 on p. 77)

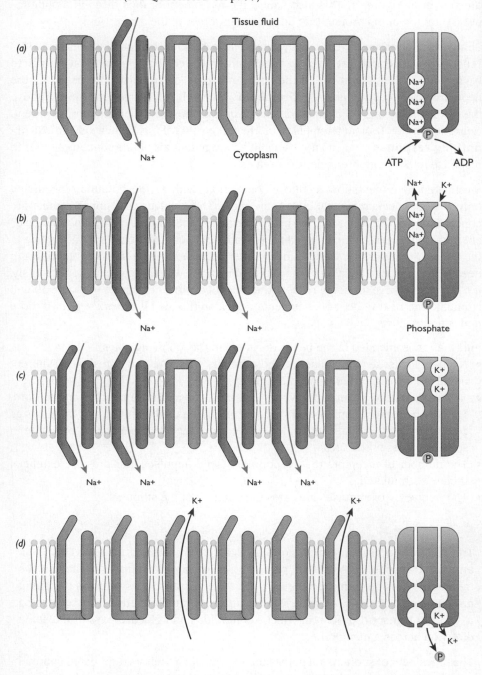

Figure 11 Ion transport at different stages of action potential: (a) during depolarisation towards threshold (b) as voltage-gated sodium ion channels begin to open in the rising phase of an action potential (c) at the height of an action potential (d) during repolarisation

Knowledge check 11

Use Figures 9 and 11 to describe the changes that occur to the voltage-gated ion channels in an axon membrane during the passage of an action potential. You may find it easiest to use a table for your answer.

A second stimulus applied to a neurone less than 1 ms after the first will not trigger another impulse. The membrane is depolarised and the neurone is in its refractory

period. Not until the –70 mV polarity is re-established will the neurone be ready to fire again. Repolarisation is established by the facilitated diffusion of potassium ions out of the neurone. In some human neurones, the refractory period lasts only 1–2 ms. This means that these neurones transmit 500–1000 impulses per second.

The events that occur in a neurone membrane when an impulse is transmitted are shown in Figure 11. This also shows the movement of ions through a sodium–potassium ion pump protein that occurs in neurones all the time.

Sometimes the stimulus received by a neurone is too low and no impulse is sent. This is because the depolarisation of the membrane of the cell body is not great enough to trigger an action potential at the axon hillock. For impulses to be sent, the depolarisation must be above a **threshold** value, which is usually 10–15 mV above the resting potential. If the depolarisation is above threshold, then an impulse is sent. Neurones either transmit impulses or they don't — this is known as the **all-or-nothing response**. They do not transmit action potentials of different amplitudes in response to different intensities of stimuli.

Examiner tip

This is also known as the all-or-none law or the all-or-nothing law.

Myelinated neurones show saltatory conduction, with action potentials occurring only at nodes, which are about 1–2 mm apart. The voltage-gated ion channels are concentrated at the nodes and are not in the membrane covered by myelin. This saves energy — fewer ion channels need to be made and pump proteins are only active at the nodes. Transmission speed is much faster because far fewer action potentials are needed in comparison with an unmyelinated neurone where they occur contiguously. If a node is inactive for some reason, then the local currents shown in Figure 10 will stimulate the next node. Local currents decay, so this will not work if two or more nodes are inactive.

Information is encoded in the nervous system in the following ways.
- Axon hillocks send impulses if the stimulus is above the threshold. Different neurones have different threshold values.
- Neurones follow distinct pathways into and out of the brain and spinal cord. Sensory neurones terminate on specific cells in the CNS that interpret impulses as coming from specific receptors. Motor neurones for specific effectors originate in specific places in the CNS.
- The number of receptors that respond and send impulses indicates the extent of the area stimulated.
- The frequency of impulses indicates the strength of the stimulus.

Examiner tip

Work through Question 2 on p. 77 to see how the strength of a stimulus influences the frequency of impulses in a neurone.

Synoptic links

The sodium–potassium ion pump is an example of active transport. This is an antiport transport protein that pumps ions in opposite directions across the membrane. It is also an example of a direct active transport system as the pump protein uses ATP. On p. 45 there is an example of indirect active transport using a symport transport protein in which sodium ions and glucose pass in the same direction across a membrane.

The movement of sodium and potassium ions through their voltage-gated channel proteins is an example of facilitated diffusion (see p. 19 of the unit guide for F211. Active transport relies on ATP made in respiration (see p. 61).

There are two common misconceptions about the role of the sodium–potassium ion pump during an action potential. One is that it stops working during the passage of the action potential and that this explains why there is a flow of ions. The other misconception is that the pump restores the resting potential. As you have seen, the resting potential is restored by the ouflow of potassium ions.

Remember that for every three sodium ions pumped out, only two potassium ions are pumped in. Therefore, for every molecule of ATP used, there is a net loss of one ion. That does not sound like much, but the cumulative effect is sufficient to reduce the concentration of ions within the cell. During the millisecond that the channels are open during an action potential, some 7000 sodium ions enter a neurone.

Synapses

Neurones make contact with their target cells at synapses. The term synapse comes from the Greek for 'fastening together' and it applies to the gap between neurone and target cell and the areas either side of the gap. In some synapses, there is no gap and impulses pass from one neurone to another electrically. We have these electrical synapses in our brains. In the synapses that we are considering, impulses are transmitted chemically using neurotransmitter substances. One of the most common neurotransmitters is **acetylcholine**. Figure 12 shows the structure of a typical synapse, which could be in the brain or spinal cord or in swellings called ganglia along some nerves. Synapses with acetylcholine as the neurotransmitter are called **cholinergic synapses**.

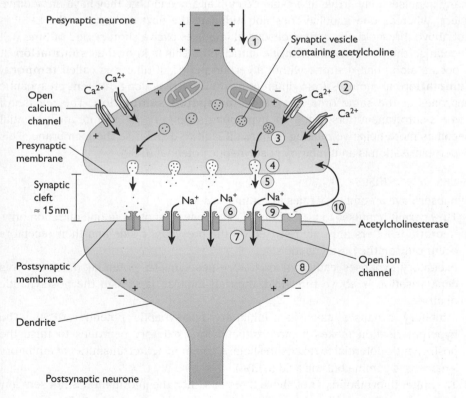

Figure 12 The events that occur during the transmission of an impulse across a cholinergic synapse

Content Guidance

(1) An action potential arrives at the end of the presynaptic neurone.

(2) Depolarisation stimulates voltage-gated calcium ion channels to open.

(3) Calcium ions diffuse through the voltage-gated calcium ion channel proteins.

(4) Calcium ions stimulate movement of synaptic vesicles towards the presynaptic membrane.

(5) Vesicles fuse with the presynaptic membrane and release acetylcholine molecules by exocytosis into the synaptic cleft. Each vesicle contains about 10 000 molecules of acetylcholine.

(6) Acetylcholine molecules diffuse across the synaptic cleft and attach to chemical-gated ion channels (this diffusion takes under 1 ms to occur).

(7) Chemical-gated ion channels open and sodium ions diffuse across the postsynaptic membrane.

(8) Influx of sodium ions depolarises the postsynaptic membrane. The depolarisation of the postsynaptic membrane is known as the excitatory postsynaptic potential (EPSP).

(9) Acetylcholine molecules leave the ion channels and enter active sites of the enzyme acetylcholinesterase, which hydrolyses each acetylcholine molecule separating the acetyl (ethanoyl) group from choline. This enzyme removes molecules of acetylcholine very quickly so stimulation is a brief event.

(10) Choline molecules enter the presynaptic membrane through transport proteins and are resynthesised into acetylcholine using energy in the form of ATP from mitochondria.

The arrival of one impulse stimulates the release of acetylcholine by exocytosis. Since many impulses may arrive at a synaptic bulb at any one time they have an additive effect. Whereas one impulse does not stimulate the next neurone as the EPSP is not above threshold, the arrival of several impulses over a short period of time will depolarise the postsynaptic neurone sufficiently. This is known as **summation**. If it occurs along one neurone within a certain period of time it is called **temporal summation**. If impulses from different neurones arrive along different presynaptic neurones at the same time it is known as **spatial summation**. The release of some neurotransmitters leads to hyperpolarisation — the membrane potential becomes more negative making it more difficult to depolarise the membrane. This hyperpolarisation is an inhibitory postsynaptic potential (IPSP).

Roles of synapses

Synapses have a number of roles in communication.

- They permit impulses to travel from one neurone to another in one direction only. Neurotransmitters are only released on the presynaptic side and their receptors occur only on the postsynaptic side.
- Excitatory synapses cause an excitatory postsynaptic potential (EPSP). If this depolarisation is above threshold, then an impulse is sent in the postsynaptic neurone.
- Inhibitory synapses cause an inhibitory postsynaptic potential (IPSP). The hyperpolarisation makes it more difficult for excitatory neurones to raise the postsynaptic potential to reach threshold. A common neurotransmitter at inhibitory synapses is gamma butyric acid (GABA).
- They filter information. If not above threshold, then the information is not sent any further. It is often *changes* in the intensity of stimuli that are detected.

Knowledge check 12

Outline what happens between the arrival of an action potential at the presynaptic membrane and the depolaristion of the postsynaptic membrane.

- Synaptic knobs run out of neurotransmitter and so develop fatigue. This stops the presynaptic neurone from stimulating the postsynaptic neurone and protects against overstimulation.
- They help to integrate information from different neurones. A single relay or motor neurone may have many thousands of neurones that form synapses over its dendrites and cell body. In this way information from many places in the body can be integrated into a single response.
- Memory is a function of the interconnections between neurones in the brain.

Synoptic links

Neurotransmitters occur in vesicles that fuse with the presynaptic membrane and are released by exocytosis. They diffuse across the synaptic gap and interact with protein receptors on the postsynaptic membrane. This is an example of cell signalling. The principles of exocytosis, diffusion and cell surface receptors are described in the guide for Unit F211 on pp. 19–20. The receptor sites on the chemically gated channel proteins are specific for the type of neurotransmitter, as is the active site of the enzyme that hydrolyses acetylcholine. These are yet more examples of the specificity of proteins.

> **Examiner tip**
>
> Many drugs work by interacting with membrane proteins at synapses. Agonists mimic the effects of the neurotransmitter; antagonists work by blocking the chemical-gated channels. There are also drugs that inhibit acetylcholinesterase. Work out the effects of these three types of drug.

Summary

- The nervous system contains neurones and glial (supporting) cells. Sensory neurones transmit impulses from receptors to the central nervous system (brain and spinal cord). Motor neurones transmit impulses from the CNS to effectors, such as muscles and glands.
- Sensory receptors are transducers; they convert different forms of energy (e.g. light, kinetic and chemical) into nerve impulses.
- The potential difference across a cell membrane is the result of unequal distribution of ions, with more anions being within the cell and more cations outside. In neurones, this is the resting potential. It is maintained by the sodium–potassium pump using energy from ATP.
- Neurones transmit nerve impulses over long distances by using reversals of the potential difference to 'boost' current flow. The opening of voltage-gated Na^+ channel proteins allows the inflow of sodium ions which depolarises the membrane from $-70\,mV$ to $+40\,mV$. The opening of voltage-gated K^+ channel proteins allows outflow of potassium ions to restore the resting potential.
- Sensory and motor neurones are protected by Schwann cells, some of which make myelin that acts as an insulator. Action potentials occur all along unmyelinated neurones. In myelinated neurones, action potentials occur at the nodes of Ranvier between Schwann cells; this gives the much faster saltatory conduction.
- Action potentials are propagated by current flow that spreads along a neurone to form local circuits that depolarise the next area of membrane to the threshold potential (e.g. $-50\,mV$) which leads to the rising phase of an action potential.
- Action potentials are generated only if the threshold for the neurone is exceeded (all-or-nothing law). The strength of a stimulus is expressed by the frequency of impulses transmitted by a neurone.
- Neurones communicate with each other at synapses. Chemical signalling occurs at synapses by release of neurotransmitter chemicals, such as acetylcholine (ACh).
- Synapses, such as cholinergic synapses, permit one-way transmission of impulses in the nervous system and interconnections between many neurones. Weak impulses are filtered out if they are below the threshold of the postsynaptic neurone; inhibitory neurones stimulate hyperpolarisation in the postsynaptic neurone making it more difficult to reach threshold.

Hormones

Key concepts you must understand

Hormones are signalling molecules that travel long distances in the blood. They are secreted by endocrine (ductless) glands. The adrenal gland (see Figure 13) is an example.

An adrenal gland is composed of an outer adrenal cortex that surrounds an inner adrenal medulla. The hormones from the adrenal cortex are steroids. Adrenaline and noradrenaline secreted by the adrenal medulla are catecholamines which are made from amino acids.

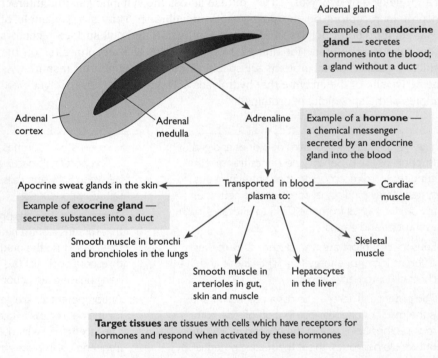

Figure 13 Adrenaline is secreted by the adrenal medulla and influences many target tissues throughout the body

Table 2 The roles of the adrenal cortex and the adrenal medulla

Region of the adrenal gland	Hormones secreted	Target organs	Functions of hormones
Adrenal cortex	Mineralocorticoid: aldosterone	Kidney, gut	Stimulates increase in absorption of sodium ions leading to an increase in blood pressure
	Glucocorticoids: cortisol and corticosterone	Liver	Stimulate increase in blood glucose concentration by gluconeogenesis
Adrenal medulla	Catecholamines: adrenaline (epinephrine*)	Liver	*Examples*: stimulate breakdown of glycogen and increase blood glucose concentration
	noradrenaline (norepinephrine*)	Heart	Increase heart rate
*Adrenaline and noradrenaline are known by international agreement as epinephrine and norepinephrine (see p. 7).			

OCR A2 Biology

Key facts you must know

Adrenaline

Adrenaline has numerous effects on the body. It is usually released in times of danger and stress and is often called the 'fight or flight' hormone. It prepares the body for meeting these situations by:

- increasing heart rate and stroke volume (volume of blood per beat)
- stimulating enzymes in the liver to convert glycogen to glucose, so releasing glucose into the blood
- decreasing blood flow to the gut and skin by stimulating vasoconstriction
- increasing blood flow to muscles and the brain by stimulating vasodilation
- increasing blood pressure because vasoconstriction increases resistance to flow
- increasing the width of bronchioles by causing smooth muscles to relax, so increasing air flow to the alveoli
- stimulating contraction of the radial muscles in the iris of the eyes to dilate the pupils
- increasing secretion from apocrine sweat glands that open into hair follicles (these glands are partly responsible for body odour)

Synoptic links

Adrenaline works together with the nervous system to increase the readiness of the body for 'flight or fight'. This includes effects on the heart (see p. 36). This is a good opportunity to revise the structure and function of the heart from Unit F211 (see pp. 41–45 of the guide for Unit F211).

As hormones circulate in the body they are broken down in the liver. Some are filtered in the kidney and excreted. This happens at different rates for different hormones. The length of time it takes for the concentration of a hormone in the blood to decrease by a half is called its half-life. Table 3 gives the half-lives of hormones mentioned in this guide.

Table 3 The half-lives of some hormones

Hormone	Half-life
Adrenaline (epinephrine); noradrenaline (norepinephrine)	160–190 seconds
Antidiuretic hormone (ADH)	10–30 minutes
Insulin	5 minutes
Glucagon	5–10 minutes
Anabolic steroids (for body building)	Between 30 minutes and 16 hours depending on the steroid

First and second messengers

Adrenaline, like insulin and glucagon (see p. 32), cannot cross cell membranes, whereas lipid-soluble steroid hormones, such as testosterone, 17 β-oestradiol and progesterone can. These steroid hormones interact with receptors inside the cytoplasm or nucleus.

Examiner tip
Cells are primed, ready to respond to the first messenger. Different types of cells are primed in different ways, which is why adrenaline stimulates very different responses in its target cells.

Cells in target tissues for adrenaline have specific **adrenergic receptors** on their cell surface membranes. Adrenaline molecules fit into these receptors and this causes a chain of events to occur inside the cell, as shown in Figure 14. The G protein acts as a transducer to transfer the message to the cytoplasm by activating the enzyme adenyl cyclase that converts ATP into **cyclic AMP**. Adrenaline is a hormone, a **first (or primary) messenger**. Cyclic AMP is a **second (or secondary) messenger** that acts on enzyme systems within the cell. It interacts with a kinase enzyme that activates other enzymes in a cascade to amplify the original message. Many enzymes are activated in the process; some are inactivated. In liver cells, the enzyme glycogen phosphorylase is activated to remove glucose units from glycogen in the form of glucose 1-phosphate molecules. These are then converted into glucose molecules, which diffuse out of the cell into the blood plasma. This increases the concentration of glucose in the blood as it is needed by muscles to contract during exercise or during 'fight or flight'. There is more about the cascade system in Question 3 on p. 81.

Examiner tip

There are other second messengers. Calcium ions play an important role in plants and animals as second messengers. The concentration in the cytoplasm is very low — almost nothing. As soon as calcium ions enter, as at a presynaptic membrane, they stimulate events to occur, such as movement of vesicles.

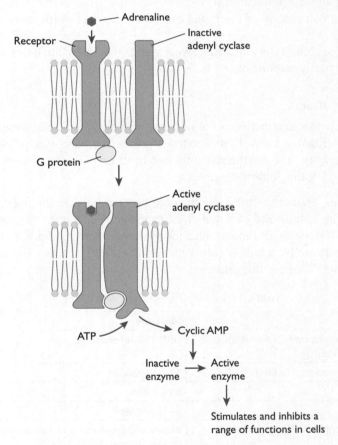

Figure 14 The role of the second messenger, cyclic AMP, inside cells stimulated by adrenaline

Focus on practical skills: histology of the pancreas

The pancreas is in the abdomen just below the stomach and to the right of the liver when viewed from the front. It is both an endocrine organ and an exocrine organ.

- Endocrine function — secretion of the hormones insulin and glucagon into the blood by cells in the islets of Langerhans; α-cells secrete glucagon; β-cells secrete insulin.
- Exocrine functions — secretion of amylase, proteases, lipase and nucleases from acinar cells into the pancreatic duct. This duct empties into the duodenum, which is the first part of the small intestine (see Figure 19 on p. 38). The alkali sodium hydrogencarbonate is also secreted to neutralise the acid contents of the stomach as they enter the duodenum.

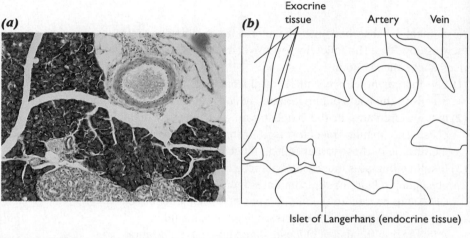

Figure 15 (a) Low-power photomicrograph of pancreatic tissue (b) A plan drawing made from the photomicrograph

Examiner tip
In the qualitative task in Unit F216, you may have to make plan diagrams and do high-power drawings from slides of the pancreas. Download images like those here and practise making such drawings. Use a microscope to look really carefully at the cells under high power.

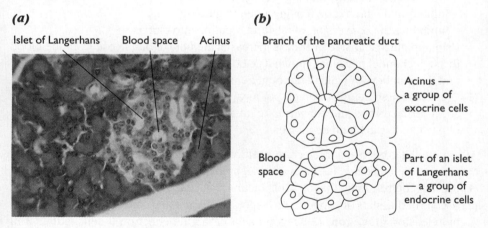

Figure 16 (a) High-power photomicrograph of the endocrine and exocrine areas of the pancreas (b) A drawing showing cells from the two areas

Knowledge check 13
Explain the difference between exocrine secretion and endocrine secretion.

Examiner tip
You should be able to interpret electron micrographs of cells from the pancreas. There are suitable electron micrographs at www.uni-mainz.de/FB/Medizin/Anatomie/workshop/EM/EMAtlas.html. At the same site there are also electron micrographs of the liver and kidney.

Controlling the concentration of blood glucose

The glucose concentration in the blood fluctuates. As blood flows through capillaries, glucose is forced out into the tissue fluid. It is important that the concentration in the tissue fluid around cells is maintained so that cells have a constant supply. Glucose is filtered out of the blood in the kidneys and is then reabsorbed so that it is

not lost in the urine (see p. 44 for more detail). If the concentration of glucose rises too high it cannot all be reabsorbed by the kidneys and *is* excreted in the urine — the concentration is said to exceed the renal threshold. This means that glucose is not stored either as glycogen or as fat and energy reserves are not topped up. If the glucose concentration falls too low, then there is not enough to supply brain cells, which cannot respire anything other than glucose, and a person may enter a coma.

The glucose concentration in the blood is kept within the range 80–120 mg per 100 cm^3 blood and is normally about 90 mg per 100 cm^3, which we can take as the set point.

As a meal is absorbed into the bloodstream the concentration of glucose may increase by 50%. The following events occur when the blood glucose concentration rises above the set point:

(1) The increasing concentration of glucose acts as a stimulus that is detected by β-cells in the islets of Langerhans, which release insulin in response.

(2) Insulin circulates in the bloodstream and binds to insulin receptors on target cells. The insulin receptor is a transmembrane protein in the cell surface membrane of these target cells in muscle, liver and adipose (fat) tissue.

(3) Insulin stimulates muscle and adipose cells to take up glucose and convert it into glycogen and fat. Insulin stimulates more channel proteins for glucose (known as GLUT4) to move into the membranes of muscle and fat cells.

(4) Insulin has a number of effects on liver cells including:
  - increasing the use of glucose, for example in respiration
  - stimulating the conversion of glucose into glycogen — this process is called glycogenesis (literally: making glycogen)
  - inhibiting the breakdown of glycogen to glucose
  - inhibiting the conversion of fats and proteins into glucose

  Using up glucose inside liver cells increases the uptake of glucose from the blood through channel proteins (known as GLUT1) that are always in the membrane and the numbers of which are not increased in response to insulin.

(5) Glucose is now being stored or 'put away' for later. It is converted to glycogen for short-term storage and converted to fat for long-term storage. The concentration of glucose in the blood decreases.

After a meal has been absorbed completely and also during exercise the blood glucose concentration may decrease below the set point. The following events then occur:

(1) β-cells stop releasing insulin. This means that cells take up less glucose.

(2) α-cells in the islets of Langerhans respond to decreasing concentrations of glucose by releasing **glucagon**. (Glucagon has a paracrine effect on β-cells, stimulating them to make insulin so they can release it when the glucose concentration increases.)

(3) Glucagon circulates in the bloodstream and binds to glucagon receptors on liver cells. The receptor interacts with adenyl cyclase to increase the concentration of cyclic AMP inside liver cells in the same way as the interaction with adrenaline (see Figure 14).

(4) Glucagon stimulates liver cells to:
- activate the enzyme glycogen phosphorylase, which converts glycogen into glucose — this process is called glycogenolysis (literally: 'splitting glycogen')
- convert fat and protein into intermediate metabolites that are converted into glucose — this process is called gluconeogenesis (literally: 'making new glucose')

Glucose accumulates inside liver cells and diffuses out into the blood, so the concentration of glucose in the blood increases.

Physiologists refer to 'the fasting state' when glucagon secretion takes place. This does not mean that a person is starving, it refers to a period of time between absorbing and assimilating one meal and the next. The fasting state refers to the time when you are asleep at night; people who have fasting blood tests are told not to eat or drink for 8–10 hours beforehand.

Figure 17 shows how the release of insulin from β-cells is controlled.

Knowledge check 14

State how the effect of insulin on liver cells differs from the effect of glucagon.

Knowledge check 15

Glucagon and glycogen are both important in the regulation of blood glucose concentration. Make a table to show five ways in which glucagon differs from glycogen.

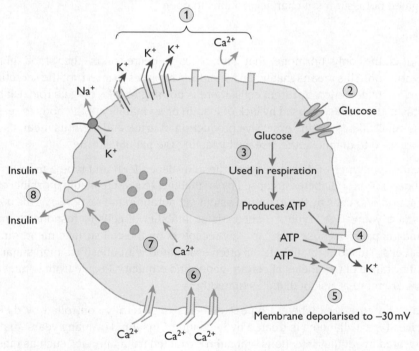

Figure 17 The release of insulin from β-cells in islet tissue in the pancreas. The numbers refer to the stages (1)–(8) described below.

(1) In the normal range of blood glucose concentration:
- membrane potential is at –70 mV; ATP-gated potassium ion channel proteins are open and some potassium ions diffuse out of the cell
- voltage-gated calcium ion channels are shut; the membrane is impermeable to calcium ions

(2) Blood glucose concentration rises as molecules from a meal are absorbed; glucose diffuses into the cell through specific channel proteins (GLUT2).

(3) Glucose is phosphorylated to glucose 6-phosphate in the first stage of respiration (see p. 62), so maintaining the concentration gradient for glucose to continue entering the cell.

(4) ATP is produced in respiration; ATP-gated potassium ion channels are sensitive to the increasing concentration of ATP and close.

(5) Potassium ions accumulate in the cell thanks to the sodium–potassium ion pump and so the positive charge *inside the cell* increases, causing a depolarisation from $-70\,mV$ to $-30\,mV$.

(6) Voltage-gated calcium channel proteins open in response to the depolarisation and calcium ions diffuse into the cell.

(7) Calcium ions stimulate the movement of vesicles towards the membrane (as they do in the synapse).

(8) Insulin is released by exocytosis and enters the blood.

Less glucose diffuses into the cell when the glucose concentration falls below the set point. Less respiration occurs and the decrease in ATP concentration causes the ATP-gated potassium ion channel proteins to open.

Diabetes

Insulin is the only hormone that stimulates a decrease in the blood glucose concentration. This means that when something happens to interrupt the secretion of insulin, or its detection by target cells, there is nothing to take over its role. Diabetes mellitus is the disease caused by lack of insulin or an inability of the body to respond to insulin. Diabetes means excessive production of urine and mellitus means sweet. Doctors used to diagnose the disease by tasting the patient's urine.

There are two types of diabetes: type 1 (insulin-dependent) and type 2 (non insulin-dependent). Type 1 diabetes is caused by an inability to secrete insulin possibly due to destruction of β-cells by the immune system (an autoimmune response). This usually starts at a young age. Type 2 diabetes is an inability of cells to respond to insulin and may happen because there are few receptors on the cell surface membranes of target cells. This form of diabetes is often associated with obesity, a high sugar diet, the inheritance of the alleles of certain genes and ethnicity. People from some ethnic groups are more at risk of diabetes than others.

There is no cure for diabetes. Type 2 diabetes is generally controlled by diet and exercise. Type 1 diabetes is treated by injection of insulin. For many years, diabetes was treated by regular injections of insulin extracted from animals, such as pigs and cattle, slaughtered for the meat trade. This form of insulin is still available, although most diabetics now receive human insulin prepared from genetically modified cells grown in fermenters.

The amino acid sequence of this form of insulin can be changed to alter the properties of the insulin. These substances are called insulin analogues. They either act faster than animal insulin (useful for taking immediately after a meal) or more slowly over a period of between 8 and 24 hours to give the background blood concentration of insulin. However, studies have not shown any advantage of using these analogues and they are more expensive, and less pure, than animal insulins. The companies that produce animal insulin have withdrawn it from many countries, but it is still available in the UK. One main advantage of GM insulin is the reliability of supply

Knowledge check 16

Explain the difference between type 1 diabetes and type 2 diabetes.

that does not depend on livestock production. With increasing numbers of people worldwide requiring insulin, this is an important consideration.

The main reported disadvantage of GM human insulin is that some people do not experience any warning signs of a hypoglycaemic attack when the blood glucose concentration falls, thus increasing the likelihood of a diabetic coma.

The problem with treating diabetes with insulin is that it has to be injected. It cannot be taken by mouth because it is a protein and would be digested in the gut by proteases. Islet cell transplantation is one possible cure. Cells taken from the islet tissue of donor pancreases are inserted into the liver where the transplanted β-cells secrete insulin. Few transplant patients have stopped taking insulin altogether, although they often need to take less insulin. However, the beneficial effects have not lasted and patients need further transplants. Stem cells are cells that retain the ability to divide by mitosis and differentiate into specialised cells, such as β-cells. Research on mice suggests that transplanted stem cells can differentiate in the host into β-cells and secrete insulin although there are problems in getting them to respond to changes in blood glucose concentrations. Rejection by the immune system is a serious problem. Putting donated islet tissue or stem cells inside porous capsules may protect them from rejection by the immune system while allowing them to secrete insulin.

Synoptic links

Glucose is transported in solution in the plasma. It is one of the components of blood that you learnt about in Unit F211. There is not much glucose in the blood or in cells at any one time — as soon as it enters cells it is converted to glucose phosphate. This ensures a steep concentration gradient for glucose. Glucose phosphate is converted into other substances, including glycogen. The advantage of the highly branched structure of glycogen is that many molecules of glucose can be released from glycogen quickly, which is useful for an animal that may have to respond within a split second and needs extra glucose in its bloodstream to be able to do so.

Insulin and glucagon are proteins. Insulin is composed of two polypeptides joined by disulfide bonds and is a good example of a protein with all four levels of organisation (see p. 11 in the guide for Unit F212). The destruction of β-cells by the immune system is an example of autoimmunity. It is not a topic that features in AS/A2 biology, but is an important class of disease that includes multiple sclerosis and rheumatoid arthritis both of which may be contexts for exam questions. You can read more about this group of diseases at:

www.nhs.uk/Conditions/Pages/bodymap.aspx

www.bbc.co.uk/health/conditions

Nervous and hormonal control

The nervous and hormonal systems act together to coordinate activities in the body. An example is the control of heart rate.

The heart is myogenic in that the stimulus for contraction originates in the heart itself in the sinoatrial node (SAN). However, the rate at which the SAN works is influenced

Examiner tip

It would be a good idea to read more about current research on type 1 and type 2 diabetes. You could start at the web sites of Diabetes UK — www.diabetes.org.uk – and Diabetes Research UK — www.jdrf.org.uk

Knowledge check 17

Explain briefly how stem cells may be used to treat diabetes.

by the accelerator nerves, which are adrenergic in that they release noradrenaline as the neurotransmitter, and the decelerator nerves that release acetylcholine. The nerves also influence the atrioventricular node (AVN). Figure 18 shows this dual innervation of the heart. (Innervation means the supply of nerves to an organ.) The heart also responds to adrenaline and noradrenaline in the blood.

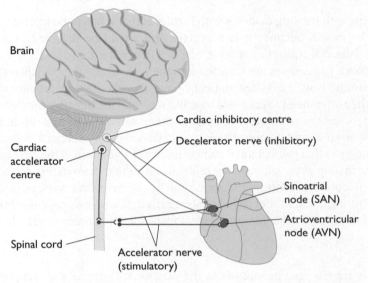

Figure 18 Dual innervation of the heart by accelerator and decelerator nerves

Table 4 Comparing control by nerves and control by hormones

Feature	Control by nerves	Control by hormones
Pathways	Along neurones	In blood plasma
Form of information	Nerve impulses	Chemicals, e.g. proteins, steroids
Speed	Fast	Slow
Target area	Cell or tissue at the end of the neurone	Whole tissues or organs; some hormones affect the whole body
Duration of effect	Short term (e.g. muscle contraction)	Short term, e.g. increase in heart rate and blood glucose concentration; long term, e.g. growth
Types of response	Muscle contraction or secretion by glands	Different responses, e.g. protein synthesis, glycogen synthesis and breakdown, vasoconstriction and vasodilation

Synoptic links

The structure and function of the heart is described in the guide for Unit F211 (see pp. 41–45). It would be useful to revise this because you may be expected to use that knowledge when answering questions on the control of the heart in the examination for this unit. Here, you should concentrate on the coordination of the heartbeat by the sinoatrial node and the atrioventricular node.

- Hormones are signalling chemicals secreted into the blood by ductless (endocrine) glands. They stimulate target tissues that are primed to make responses.

- The cortex of adrenal glands secretes steroid hormones to regulate aspects of metabolism and ion uptake in the kidney. The medulla secretes adrenaline, which affects many target tissues during responses to emergencies.

- Adrenaline is a first messenger that interacts with receptors on the cell surface membrane of target cells. In response, cells make cyclic AMP which is a second messenger that activates the first enzyme in a cascade of enzymes to amplify the message for a fast response.

- Exocrine cells in the pancreas secrete digestive enzymes into ducts that empty into the duodenum. The islets of Langerhans are groups of endocrine cells scattered throughout the exocrine tissue: α-cells secrete glucagon and β-cells secrete insulin to regulate blood glucose concentration.

- Insulin stimulates liver and muscle cells to store glucose in the form of glycogen, so reducing blood glucose concentration. Glucagon stimulates liver cells to breakdown glycogen to glucose, so increasing blood glucose concentration.

- Insulin secretion is controlled by ATP-gated potassium ion channels in β-cells. They close in response to an increase in blood glucose concentration above the set point. The cells depolarise and voltage-gated calcium ion channels open. Calcium ions enter and stimulate movement of vesicles to release insulin by exocytosis.

- Diabetes mellitus is a disease in which the homeostatic control of blood glucose fails to function. In type-1 diabetes, β-cells fail to release insulin. Type 2 diabetes begins later in life and is a failure of cells to respond to insulin. Type-1 diabetes is treated with injections of insulin; type-2 by other drugs, careful diet and exercise.

- The production of insulin by GM cells allows a constant supply of human insulin, rather than a supply dependent on the meat trade in providing animal insulin. The transplant of stem cells that differentiate into β-cells may be a long-term treatment or even a cure.

- Heart rate is controlled by neurones in the cardiac accelerator and decelerator nerves. Adrenaline also stimulates an increase in heart rate.

Module 2: Excretion

Key concepts you must understand

Excretion is the removal from the body of toxic waste products of metabolism and substances that are in excess of requirements. Metabolic waste products are:

- carbon dioxide from respiration, which is lost through the gas exchange system
- nitrogenous waste in the form of ammonia, urea and uric acid

Carbon dioxide is produced by decarboxylation of respiratory substrates (see p. 64). Ammonia is produced by the deamination of excess amino acids. If allowed to accumulate, both carbon dioxide and ammonia would change the pH of cytoplasm and body fluids and this would cause enzymes to work less efficiently.

The excretory system consists of the liver, which produces many of the excretory products, and the kidneys, ureters, bladder and urethra. The lungs remove most of the carbon dioxide, so can also be considered part of the excretory system.

Key facts you must know

The liver

The liver is a large organ situated in the abdomen to the left of the stomach, when viewed from the front. It carries out many functions including roles in temperature

Examiner tip
If all the carbon dioxide from respiring tissues dissolved in the plasma, the pH would decrease significantly and death would occur in a very short time. In F211, you studied the ways in which carbon dioxide is transported that avoid fluctuations in blood pH.

regulation (see p. 13), regulation of metabolism, protein synthesis, transamination, synthesis of cholesterol and bile salts, digestion, assimilation and excretion.

Focus on practical skills: histology of the liver

Figures 19, 20 and 21 show the structure of the liver.

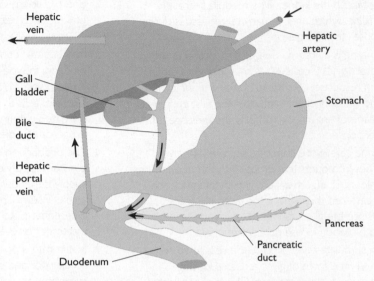

Hepatic vein

Hepatic artery

Gall bladder

Bile duct

Hepatic portal vein

Stomach

Pancreas

Pancreatic duct

Duodenum

Figure 19 The gross structure of the liver, its blood supply and connection to the duodenum

The liver is divided into lobules. Figure 20 shows that the lobules are separated by areas of connective tissue. Oxygenated blood flows into the lobules in branches of the hepatic artery (itself a branch of the aorta) and deoxygenated blood rich in absorbed food from the digestive system enters in branches of the hepatic portal vein. This blood flows through wide capillaries known as sinusoids that are lined by an incomplete layer of endothelial cells, which allows the blood to reach the hepatocytes and makes exchange of substances between blood and cells easy.

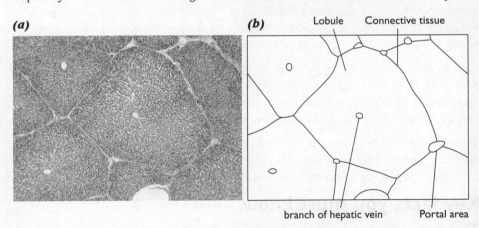

(a)

(b) Lobule Connective tissue

branch of hepatic vein   Portal area

Figure 20 (a) A low-power photomicrograph of some liver lobules (b) A plan drawing made from the photomicrograph

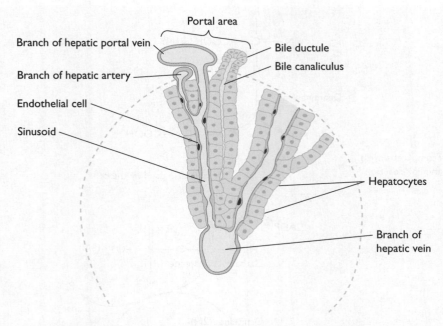

Figure 21 Arrangement of blood vessels, sinusoids, hepatocytes and bile canaliculi inside each liver lobule

Hepatocytes make bile, which is a digestive secretion that is stored in the gall bladder and enters the duodenum. Bile contains bile pigments, which are excretory products made from haemoglobin, and bile salts for the emulsification of fats. These substances enter the little channels (canaliculi) which join together to form bile ducts that drain into the gall bladder and the duodenum. The blood flows along the sinusoids from the branches of the hepatic artery and the hepatic portal vein to drain into a branch of the hepatic vein. This deoxygenated blood flows back to the heart through the vena cava.

Formation of urea

The body requires amino acids to make proteins. At any one time, the intake of amino acids may exceed the requirements. Excess amino acids cannot be stored and are broken down by enzymes in the liver. The amino group from each amino acid is removed to form ammonia (NH_3) and an organic acid in **deamination**. The organic acid is used in respiration or in the synthesis of other compounds.

Ammonia is made less harmful by a series of reactions that occur in liver cells. These reactions form a cycle — the **ornithine** (or **urea**) **cycle**. Figure 22 shows deamination and an outline of the ornithine cycle. On the diagram, the numbers of nitrogen atoms in each compound are shown in brackets. The production of carbamyl phosphate and citrulline occurs in mitochondria; the other reactions occur in the cytosol. Aspartate is formed from excess amino acids.

Urea is soluble in water and is less toxic than ammonia. It diffuses readily through membranes and so leaves hepatocytes and is transported to the kidneys dissolved in the blood plasma. We also excrete tiny quantities of ammonia and uric acid (see Question 4 on p. 85).

(see Question 4 on p. 85).

Examiner tip

The ornithine cycle results in the production of one molecule of urea from two amino groups and one molecule of carbon dioxide. The advantage of a cycle is that only small quantities of the intermediates are needed to process large quantities of substances, such as amino groups and carbon dioxide.

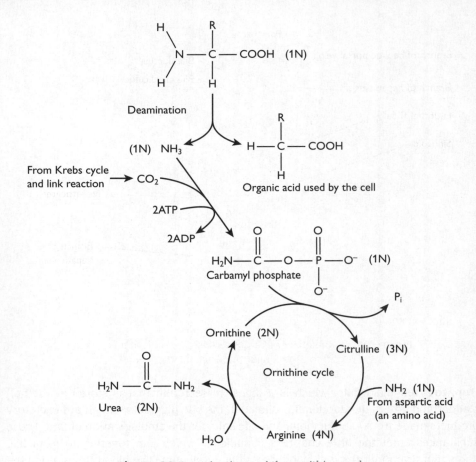

Figure 22 Deamination and the ornithine cycle

Examiner tip
Make a labelled diagram of a mitochondrion. Annotate your diagram with the processes that occur within it. You can add more annotations as you read the rest of this guide. Use the diagrams on pp. 64–65 to give further details of the structure of this organelle.

Coenzymes

Coenzymes are organic cofactors that take part in enzyme-catalysed reactions and as a result are changed slightly. Coenzyme A plays a central role in metabolism. It is present in mitochondria and receives small 2-carbon compounds from the breakdown of carbohydrate, some amino acids and fatty acids. It passes these 2-carbon compounds to the reactions of the Krebs cycle for further metabolism. The transfer of hydrogen atoms is an important part of many metabolic pathways. Coenzymes NAD, FAD and NADP are hydrogen carriers that are reduced when they receive a hydrogen atom(s) and oxidised when hydrogen is passed on. These coenzymes transfer hydrogen in a controlled fashion via different enzymes.

There are different ways of representing the reduction of NAD. In this unit it is acceptable to write NAD (for the oxidised form) and reduced NAD, without including the hydrogen atoms. If you are taking chemistry, you may find this unacceptable, in which case you can write:

$$NAD^+ \quad + 2H \rightarrow \quad NADH \quad + H^+$$
$$\text{Oxidised form} \qquad \text{Reduced form}$$

In the reduced form, a proton (H^+) is associated with the NAD molecule. NAD is a mobile coenzyme and shuttles between enzymes in different parts of the cell.

OCR A2 Biology

NAD and FAD are involved in reactions in respiration. NADP is involved in the reactions that occur in photosynthesis (see p. 54). The changes to NADP are written in the same way as those for NAD.

You may also refer to FAD (for the oxidised form) and reduced FAD. FAD is not mobile — it is tightly bound to an enzyme and is that enzyme's prosthetic group. When reduced, FAD forms $FADH_2$ and the reaction is written as:

$$FAD \quad + 2H \rightarrow \quad FADH_2$$
Oxidised form   Reduced form

ATP also acts as a coenzyme (there is more detail about ATP on p. 60).

Detoxification

The liver breaks down toxic substances, such as alcohol. This is known as **detoxification**.

The metabolism of alcohol is linked with the reduction of NAD, which is recycled by oxidation in mitochondria. This in turn generates ATP for the cell. Since the metabolism of alcohol generates plenty of energy as ATP, liver cells do not use as much fat as usual, so this is stored within the cells giving rise to the condition 'fatty liver'. The fat stored in the liver reduces the efficiency of the hepatocytes in carrying out their many functions. Fatty liver can lead to life-threatening conditions such as cirrhosis, a condition that is on the increase in the UK among young people who misuse alcohol.

Other drugs, such as paracetamol, steroids and antibiotics are also broken down in the liver. Hepatocytes contain catalase, the enzyme that breaks down hydrogen peroxide, which is a toxic end product of metabolism.

Synoptic links

The structure of amino acids was covered in Unit F212. Liver cells make albumen which is the major plasma protein that maintains the water potential of blood plasma.

This would be a good opportunity to revise the structure and formation of peptide bonds in the synthesis of proteins (see p. 10 of the guide for Unit F212; details of protein synthesis are in Unit F215).

This unit covers metabolic pathways involving the biochemical substances you studied in Unit F212. The ornithine cycle is the first of these. Later, you will learn about photosynthesis and respiration, which have linear, branched and cyclical pathways.

FAD, NAD and NADP are coenzymes that are composed of nucleotides. There are a number of other coenzymes with similar structures that are involved in many metabolic pathways. See p. 32 of the guide for Unit F212 for information about cofactors and coenzymes.

The kidneys

The kidneys are the main excretory organs situated at the back and top of the abdomen just below the diaphragm. The kidneys filter blood and excrete waste

> **Examiner tip**
> NADP is the coenzyme involved in photosynthesis. You can remember the P as standing for photosynthesis, although in fact it refers to phosphate. NADP is not only found in plants; we use it in metabolic pathways in lipid and nucleotide synthesis.

products and substances in excess of requirements. They are the effector organs in water regulation.

At birth, each kidney contains about a million unit structures known as nephrons. The number decreases with age. Each nephron filters blood to produce a liquid (filtrate) from the plasma, which contains useful substances as well as waste substances. Nephrons reabsorb useful substances into the blood and control the volume of water lost in the urine.

Focus on practical skills: histology of the kidney

Figures 23, 24 and 25 show the structure of the kidney.

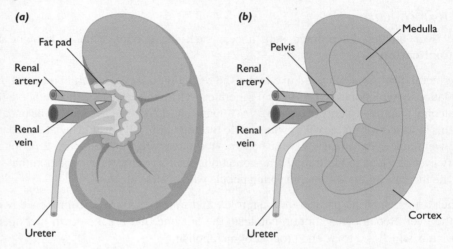

Figure 23 (a) The gross structure of a kidney: external view (b) A vertical section

Most textbook diagrams of vertical sections of the kidney show the position of at least one nephron. However, it is not possible to see complete nephrons in a kidney section because they do not lie in any one plane within the kidney. Sections like those in Figures 24 and 25 show the parts of a number of adjacent nephrons.

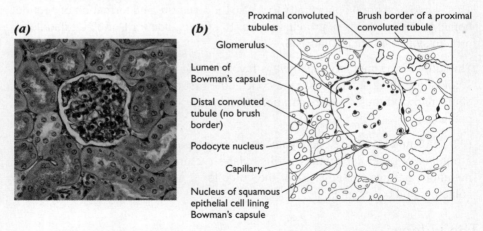

Figure 24 (a) A photomicrograph of part of the cortex of the kidney (b) A drawing of part of the photomicrograph

The outer part of the kidney is the cortex. It contains glomeruli, proximal convoluted tubules and distal convoluted tubules. Each glomerulus consists of a tightly arranged group of capillaries. These capillaries sit inside a cup-like structure called the Bowman's capsule (also known as the renal capsule).

(a)　　　　　　　　　　**(b)**

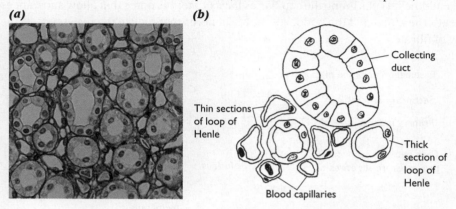

Collecting duct

Thin sections of loop of Henle

Thick section of loop of Henle

Blood capillaries

Figure 25 (a) A photomicrograph of part of the medulla of the kidney (b) A drawing of part of the photomicrograph

The inner part of the kidney is the pelvis where urine collects. Between it and the cortex is the medulla which contains loops and collecting ducts. These are visible in Figure 25 in cross-section.

Figure 26 shows the structure of a single nephron and its associated blood vessels.

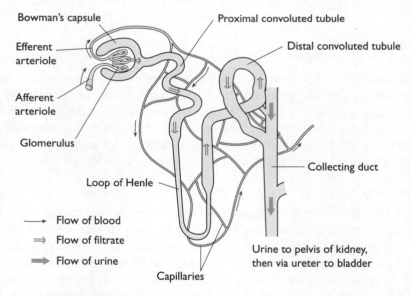

Bowman's capsule

Proximal convoluted tubule

Efferent arteriole

Distal convoluted tubule

Afferent arteriole

Glomerulus

Collecting duct

Loop of Henle

→ Flow of blood
⇒ Flow of filtrate
⇒ Flow of urine

Urine to pelvis of kidney, then via ureter to bladder

Capillaries

Figure 26 A kidney nephron and associated blood vessels

Ultrafiltration

Blood enters the **glomerulus** from a branch of the renal artery at high pressure. Notice that the diameter of the efferent arteriole is narrower than that of the afferent

Examiner tip

Afferent and efferent are words used in anatomy to describe structures, such as blood vessels, that 'go to' or 'come from' an organ or other structure. You need to use these words when explaining ultrafiltration. Remember that **e**fferent equals **e**xit.

arteriole. This builds up a 'head of pressure', which forces molecules with a relative molecular mass of less than 69 000 through the glomerular capillaries into the Bowman's capsule. This **pressure filtration** occurs in all capillaries, but is made more effective here by the arrangement of capillaries and podocytes (see Figure 27). The endothelial cells lining the capillaries have numerous pores that allow substances to leave the blood and the podocytes have slit pores that reduce the resistance to the flow of filtrate.

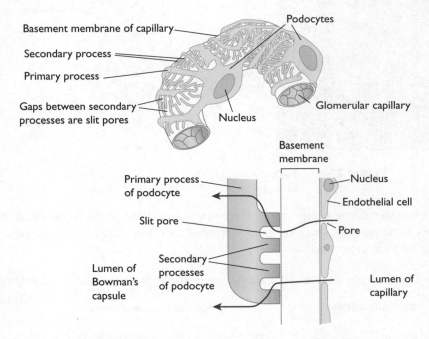

Figure 27 Ultrafiltration in the glomerulus

Knowledge check 18

The volume of blood that passes through the glomeruli is 1200 cm³ per minute. Of this, 125 cm³ forms filtrate. The plasma forms 55% of the blood by volume. Calculate the percentage of the plasma that forms filtrate every minute.

On the outside of the endothelial cells is a basement membrane made of glycoproteins. This acts as a molecular sieve, retaining large molecules, such as proteins, and allowing small molecules to pass through. The capillaries in the glomerulus are supported by podocytes that form an incomplete layer so they do not offer resistance to the flow of filtrate. The processes of the podocytes are arranged like loosely interlocked fingers. The spaces between the fingers represent the slit pores.

Selective reabsorption

The filtrate collects in the Bowman's capsule and passes into the proximal convoluted tubule (PCT). The cells that line the PCT are specialised for reabsorption of useful substances from the filtrate. Figure 28 shows the structure of these cells and how they reabsorb sodium ions and glucose (amino acids are reabsorbed by similar methods). Urea diffuses across the cells back into the bloodstream. The movement of solutes from the filtrate to the blood gives the blood a lower water potential than the filtrate. As a result, water passes by osmosis from the filtrate back into the blood. The reabsorption of urea and water is passive.

Absorption of glucose is an active process that requires a supply of ATP from respiration. ATP provides energy for sodium–potassium ion pump proteins on the lateral and basal membranes of the cells. The pumps create a low concentration of sodium ions inside the cytoplasm. This means there is a concentration gradient for sodium ions from the filtrate into the cytoplasm, which is used to drive the uptake of other molecules, such as glucose and amino acids. As we have seen, sodium ions can only diffuse through specialised protein channels or carriers. The protein carrier on the luminal membrane facing the filtrate is a **symport** (co-transporter protein) that has binding sites for sodium ions and glucose. When both of these are filled, the carrier changes shape to deliver the sodium ion and glucose into the cytoplasm. This gives a high concentration of glucose inside the cell and glucose molecules diffuse out through the basal and lateral membranes into the blood. The absorption of glucose and amino acids in this way is an example of indirect active transport — the molecules move into and out of the cell but this movement is driven by the active transport of sodium ions and the presence of the co-transporter protein in the luminal membrane.

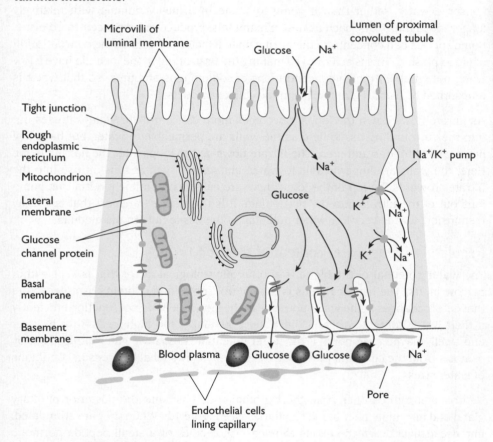

Figure 28 Selective reabsorption by a cell of the proximal convoluted tubule

PCT cells are a good example of the relationship between structure and function:
- Tight junctions (like Velcro®) between cells ensure movement occurs through the cells, not between them.
- Microvilli give a large surface area for many symport carrier proteins.
- Mitochondria produce ATP for active transport by Na^+/K^+ pumps.
- Infoldings of the basal membrane give a large surface area for movement of substances into the blood.

- Rough endoplasmic reticulum makes proteins for Na^+/K^+ pumps, symport carrier proteins and glucose channel proteins.

By the time the filtrate has reached the end of the PCT a large proportion of the solutes and much of the water has been reabsorbed. The remaining filtrate passes into the loop and then into the distal convoluted tubule (DCT).

Creating a water potential gradient

The main function of the rest of the nephron is to regulate the concentration of the blood and determine the concentration of the urine. The maximum concentration of urine in mammals is determined by the lengths of the loops and therefore by the width of the medulla. Humans are able to concentrate urine by a factor of four compared with the concentration of blood plasma. This means that when dehydrating we can conserve water, rather than it going to waste in urine. Mammals with relatively longer loops than ours, such as desert mammals, produce more concentrated urine. Some species can concentrate their urine up to a factor of 25 times the concentration of blood plasma. This is achieved by making the tissue fluid in the medulla have a low water potential (with high concentrations of sodium ions and urea) so that water is reabsorbed by osmosis from urine in the collecting ducts.

As filtrate flows down the loops it becomes more concentrated. Water diffuses out into the surrounding tissue fluid as the walls are permeable to water, but have low permeability to ions and urea. The filtrate flows up the loops after the hairpin bend. Here, the walls are impermeable to water, but solutes such as ions diffuse into the filtrate. Towards the top of the loops there are cells rich in mitochondria that pump ions out of the filtrate into the tissue fluid. It is this active transport that is mainly responsible for the very low water potential of the tissue fluid in the medulla.

Regulation of the water content of the blood

The water potential of the blood is another physiological factor that is kept within narrow limits. The hypothalamus is the control centre and contains osmoreceptors that detect changes in the water potential of the blood. Cells do not function efficiently if their water content varies. If the water potential is too high, cells absorb water and swell, maybe even burst. If the water potential is too low, water diffuses out by osmosis from the cells and they shrink. Figure 29 shows what happens in conditions of water stress.

Neurosecretory neurones from the hypothalamus pass into the posterior pituitary gland and terminate near blood capillaries (see Figure 1d). When they are stimulated, impulses travel down the axons to release molecules of a small peptide hormone called antidiuretic hormone (ADH) by exocytosis of vesicles. The target cells of ADH are the cells of the DCT and collecting ducts. Figure 30 shows what happens when ADH attaches to receptors on the surface membranes of these cells. Aquaporins are water channels that become inserted in the cell-surface membrane. When open, 3 billion molecules of water a second move through each aquaporin. (There are many different types of aquaporin; the type that responds to ADH is aquaporin 2.)

Knowledge check 19

State the changes to the filtrate as it flows through the PCT.

Examiner tip

Make a large, simple diagram of the nephron with a long loop of Henle and a collecting duct. Draw in the blood vessels in the medulla as a loop. Note that loops of Henle, blood vessels (known as the vasa recta) and collecting ducts are all parallel.

Knowledge check 20

Aquaporins are selective for water and do not allow ions to pass through. Suggest how this happens.

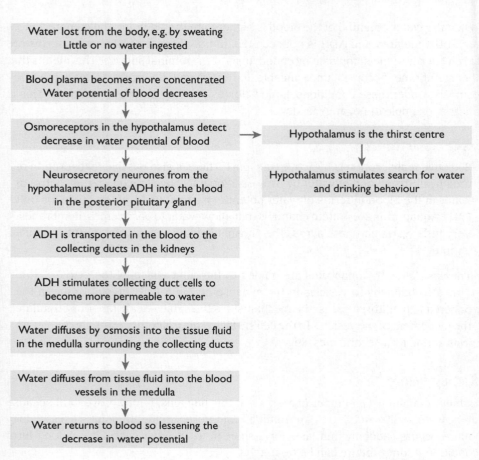

Water lost from the body, e.g. by sweating
Little or no water ingested

↓

Blood plasma becomes more concentrated
Water potential of blood decreases

↓

Osmoreceptors in the hypothalamus detect decrease in water potential of blood → Hypothalamus is the thirst centre

↓

Neurosecretory neurones from the hypothalamus release ADH into the blood in the posterior pituitary gland

Hypothalamus stimulates search for water and drinking behaviour

↓

ADH is transported in the blood to the collecting ducts in the kidneys

↓

ADH stimulates collecting duct cells to become more permeable to water

↓

Water diffuses by osmosis into the tissue fluid in the medulla surrounding the collecting ducts

↓

Water diffuses from tissue fluid into the blood vessels in the medulla

↓

Water returns to blood so lessening the decrease in water potential

Figure 29 Involvement of the kidney in osmoregulation

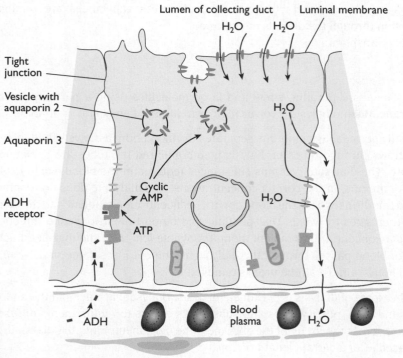

Figure 30 The changes that occur in a cell in a collecting duct in response to stimulation by ADH (aquaporin 1 is found in cell surface membranes of the PCT and descending limb of the loop of Henle)

Examiner tip
Note that in the USA, ADH is known as vasopressin and you will see this term used on websites.

When the water potential of the blood is higher than the set point, the osmoreceptors are not stimulated and ADH is not secreted. In the absence of ADH, aquaporins are taken back into the cytoplasm by endocytosis at the luminal surface. This means that the membrane becomes impermeable to water and no water is reabsorbed by osmosis. Under these conditions, large volumes of dilute urine are produced. This is another example of negative feedback.

Synoptic links

Selective reabsorption is by active transport. There is also passive reabsorption of urea by diffusion and of water by osmosis. You may have to explain the control of water in the blood in terms of water potential (see pp. 19–21 of the guide for Unit F211). Aquaporins are protein channels that allow water to pass across membranes. Very little water can pass across the phospholipid bilayer, so aquaporins provide a route.

The vesicles with aquaporins are guided to the cell surface by the cytoskeleton. This also happens to vesicles in the synapse and in β-cells in islet tissue in the pancreas. In all three cases, the mechanism is the same — calcium ions stimulate the movement of the vesicles to the cell surface. This is another example of calcium ions acting as a second messenger.

Kidney failure

Kidneys can fail for a number of reasons. If this happens, it may prove fatal because urea, water, salts and toxins are retained and not excreted. In some cases, kidney failure occurs suddenly and lasts for a short time. In other cases, it is a long-term condition. Kidney failure can be treated by:
- dialysis — toxins, metabolic wastes and excess substances are removed by diffusion through the dialysis membrane
- kidney transplant

Renal dialysis

Dialysis means separating small and large molecules using a partially permeable membrane. Haemodialysis involves regular treatment in hospital or at home.

A partially permeable membrane separates the blood from a dialysis fluid (dialysate). Blood flows through tubes of dialysis membrane and the tubes are surrounded by dialysate. The dialysate contains substances required in the blood, such as glucose and sodium ions, in the correct concentrations. The dialysate does not contain any urea, so this diffuses from the blood down its concentration gradient into the dialysate, which then goes to waste. Dialysis fluid and blood flow in opposite directions so there is a concentration gradient along the whole length of the dialyser. Each time a unit of blood passes through the dialysis machine it loses some of its urea. After several hours, almost all the urea is removed.

The other form of dialysis is peritoneal dialysis. Dialysate is placed in the abdominal cavity, and urea, other wastes and substances excess to requirements diffuse from the blood across the lining of the abdomen (the peritoneum) into the dialysate which is replaced after a certain length of time.

Dialysis places severe restrictions on patients' lives. They may have to make regular trips to a hospital or clinic to receive treatment. Diet has to be controlled carefully so that they do not produce too much urea or ingest too much salt. A solution to these problems is to carry out a kidney transplant. Only one kidney is required, but the problem is finding a donor. The blood groups of donor and recipient must be compatible and there also needs to be a reasonably good match between the tissue types.

Urine tests

Urine samples are used in pregnancy tests. The testing sticks contain molecules of a monoclonal antibody specific to the hormone human chorionic gonadotrophin (hCG), which is secreted by the early embryo shortly after implantation. The monoclonal antibodies used in pregnancy testing are manufactured by culturing modified B-lymphocytes that secrete antibodies which have antigen binding sites complementary to the hormone hCG.

Athletes are tested regularly to check that they have not been taking anabolic steroids to build up muscle mass (anabolic steroids stimulate protein synthesis). They are detected in the urine by gas chromatography or mass spectrometry. These drugs are based on steroid hormones and may have a half-life of 16 hours so it is important that samples are taken close to the event (see Table 3, p. 29).

Knowledge check 21

Insulin is detected in the urine. Suggest why this is so.

Synoptic links

Filtrate is similar to tissue fluid in that it has been filtered from the blood by ultrafiltration through the walls of capillaries. It is formed from plasma. Urine is formed from filtrate and has a different composition. In Unit F211 you studied the differences between plasma, tissue fluid and lymph. You could draw up a table to show the components of these three body fluids and urine to show which components occur in each fluid. You could use the table in Question 4 on p. 85 to help you.

In the UK, kidneys for transplant are usually matched exactly for blood group. You may already know that the ABO blood group is controlled by a gene with codominant alleles. This is covered in Unit 215. Kidneys are rejected because of a mismatch in tissue types — it is rarely possible to match these exactly. The immune system of the recipient detects the antigens on the surface of the kidney cells as being foreign (non-self). This stimulates an immune response resulting in rejection of the kidney. People who have had a kidney transplant take immunosuppressive drugs to prevent the immune system from destroying the kidney. The immune response here is complex because it involves the immune system destroying a large number of foreign cells. T-lymphocytes play an important role in detecting and destroying the cells (see pp. 46–47 of the guide for Unit F212).

- Excretion is the removal of metabolic waste and substances that are in excess of requirements.

- The liver receives blood from the hepatic artery and the hepatic portal vein; it is drained by the hepatic vein. Bile is stored in the gall bladder and passes to the duodenum in the bile duct. Liver cells (hepatocytes) are arranged into lobules in which each cell is in contact with blood.

- Excess amino acids are deaminated: the $-NH_2$ group forms ammonia, which is toxic. The reactions of the ornithine cycle combine ammonia and carbon dioxide to form the less toxic urea.

- The liver detoxifies drugs, such as alcohol, paracetamol and antibiotics.

- The kidneys are the main excretory organs. Blood is supplied in the renal artery and drained in the renal vein. Urine flows in the ureter to the bladder.

- The functional unit of the kidney is the nephron. Ultrafiltration occurs in the glomerulus; filtrate collects in the Bowman's capsule; selective reabsorption by active transport, diffusion and osmosis occurs in the proximal and distal convoluted tubules. The loops of Henle and collecting ducts maintain a high concentration of solutes in the tissue fluid in the medulla.

- The cortex is the outer region of the kidney; it surrounds the inner medulla and pelvis. The cortex contains numerous glomeruli; the medulla contains many loops, blood vessels and collecting ducts that run in parallel.

- Osmoreceptors in the hypothalamus detect changes in the water potential of plasma. When plasma becomes too concentrated, ADH is secreted by the posterior pituitary gland. Cells in the distal tubule and collecting ducts respond by inserting aquaporins in their cell membranes; water diffuses through these into tissue fluid in the medulla to conserve water.

- Treatment for people with kidney failure is regular dialysis or a kidney transplant.

- The hormone hCG is secreted from the very beginning of pregnancy and excreted in urine. Pregnancy tests involve using monoclonal antibodies to detect hCG. Other hormones, such as anabolic steroids, which are misused by some athletes, are also detected in urine samples.

Module 3: Photosynthesis

Key concepts you must understand

Energy flows in ecosystems from the Sun through plants, animals and decomposers (bacteria and fungi). Organisms respire much of the energy that they take in and transfer much of that energy as heat to their surroundings. Producers, such as green plants and algae, convert light energy to chemical energy by photosynthesis, so making energy available to all other organisms in most ecosystems. Producers use light energy to convert simple inorganic molecules to complex organic molecules. They are **autotrophs**. Animals and decomposers cannot convert simple molecules to complex molecules. They obtain energy in the form of complex organic compounds and are known as **heterotrophs**.

Photosynthesis provides the complex carbon compounds that are the basis of the metabolism of plants and animals. Respiration is the metabolic process that transfers energy from complex carbon compounds to ATP, making energy available for cell activities (see p. 60). All organisms carry out respiration. Plants gain the carbon compounds they require from photosynthesis. Animals gain their carbon by eating plants or other animals. Decomposers gain carbon compounds from dead plant and animal material.

Key facts you must know

Photosynthesis is a two-stage process that occurs entirely inside chloroplasts. The two stages are the **light-dependent stage** and the **light-independent stage**.

Figure 31 shows the structure of a chloroplast and the exchanges that occur between a chloroplast and its surroundings. Figure 32 shows the relationships between the two stages of photosynthesis.

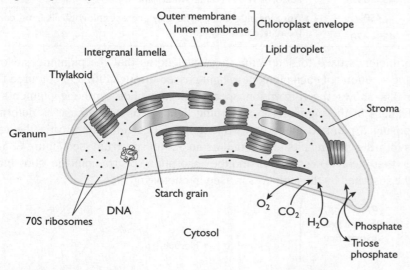

Figure 31 The structure of a chloroplast and the exchanges that occur with the cytosol

Examiner tip

You studied chloroplasts in Unit F211. You should be able to recognise the different parts of this organelle in electron micrographs. Have a look at some to prepare for labelling the parts of a chloroplast in an exam.

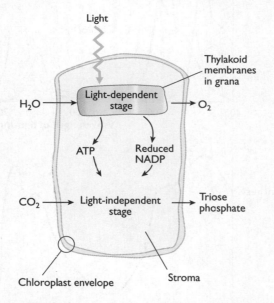

Figure 32 The location of the two stages of photosynthesis

Photosynthetic pigments

A pigment is a coloured compound that absorbs some wavelengths of light and reflects others. The colour we see is the colour that is reflected. A photosynthetic pigment is a coloured compound that absorbs light energy over a range of wavelengths with a peak at a certain wavelength. Photosynthetic pigments can be separated by chromatography.

Table 5 Photosynthetic pigments

Pigment	Colour	Peak absorption/nm	Function in photosynthesis
Chlorophyll a	Yellow-green	430, 662	Absorbs red and blue-violet light
Chlorophyll b	Blue-green	453, 642	Absorbs red and blue-violet light
β-carotene	Orange	450	Absorbs blue-violet light; may protect chlorophylls from damage from light and oxygen
Xanthophylls	Yellow	450–470	

A colorimeter is used to determine the wavelengths that the pigments absorb by exposing solutions of each pigment to light passed through different coloured filters. Figure 33a shows the absorption spectra for the photosynthetic pigments. The effectiveness of the pigments in absorbing light for photosynthesis is determined by exposing a plant to different wavelengths of light and measuring the rate of photosynthesis. The results are plotted as an action spectrum (see Figure 33b). You can see that the action spectrum and the absorption spectra coincide, showing that the light absorbed by the pigments is used in photosynthesis.

Figure 33 (a) Absorption spectra for photosynthetic pigments and (b) the action spectrum for photosynthesis

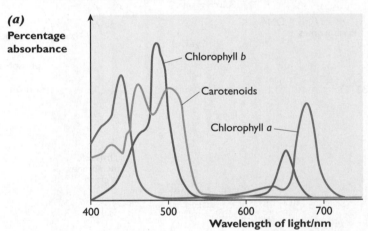

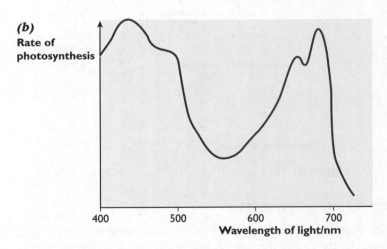

The light-dependent stage

There are some good animations of the light-dependent stage — for example:

http://highered.mcgraw-hill.com/sites/0072437316/student_view0/chapter10/animations.html

www.johnkyrk.com/photosynthesis.html

This stage occurs on the thylakoid membranes. Light is absorbed, providing energy to split water. The products of the light-dependent stage are ATP and reduced NADP. The process involves the transport of electrons through a series of compounds linked to the movement of hydrogen ions (protons) across the thylakoid membranes.

Two photosystems, PSI and PSII, are situated in the membranes. These consist of light-harvesting complexes (LHCs) with different pigments. Two types of chlorophyll *a* molecule form the reaction centres of the photosystems. Chlorophyll *a* with peak absorption at 700 nm is the reaction centre of PSI; chlorophyll *a* with peak absorption at 680 nm is the reaction centre of PSII. Pigments in the LHCs absorb photons of light and pass the energy to the reaction centres, P700 and P680.

Light energy excites electrons in P680 in PSII so that they gain enough energy to leave the molecule and pass to a series of protein carrier molecules. As electrons move through the chain (known as the electron transport chain or ETC) they lose energy which is harnessed to move protons by a form of active transport across the thylakoid membrane into the thylakoid space. The electrons then pass to PSI where they are re-energised by the absorption of light energy and pass through more carriers to reduce NADP on the stromal side of the membrane.

Proton pumping results in a higher concentration of protons in the thylakoid space than in the stroma. The protons can only leave the thylakoid space through ATP synthase. The diffusion of protons down their concentration gradient through ATP synthase provides energy for the synthesis of ATP from ADP and phosphate ions. The use of light energy to drive the production of ATP is **photophosphorylation**.

Water provides protons for photophosphorylation, and protons and electrons for the reduction of the coenzyme NADP. In the thylakoids, there is a water-splitting enzyme that breaks down water to give hydrogen ions, free electrons and oxygen:

$$2H_2O \rightarrow 4H^+ + 4e^- + O_2$$

The light-dependent stage can be shown as the Z-scheme. This shows the route followed by electrons as they pass from water to NADP in the thylakoid membranes. It is actually a Z rotated 90° to the right and looks more like the letter N. It is a graph in which the vertical axis is the energy level of electrons at different stages of the pathway. In the vertical part of the N, electrons gain energy from light. In the falling part of the N, electrons lose energy, which is used to drive protons into the thylakoid space. Figure 34 does *not* show the Z scheme — it shows the interaction between the components of the light-dependent stage and the positions they occupy in the thylakoids.

> **Examiner tip**
> Z-scheme diagrams have a vertical scale that shows the ability of each molecule to transfer an electron to the next in the chain. Energy is available to pump protons in the 'downhill' sections. Figure 34 shows the position of the molecules in the chloroplast, not their 'energy values'.

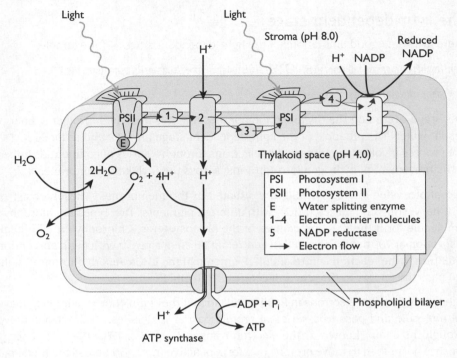

Figure 34 The light-dependent stage of photosynthesis

The sequence of processes in the light-dependent stage is as follows:

- Light energy is absorbed by LHCs.
- Water is split by photolysis to give H^+ and electrons.
- Oxygen is released.
- Energy is harnessed when electrons flow along chains of electron carriers by active transport of protons (proton pumping). A proton gradient is formed.
- ATP is formed in:
    - cyclic photophosphorylation
    - non-cyclic photophosphorylation
- In non-cyclic photophosphorylation, reduced NADP is formed.

The flow of electrons in Figure 34 is non-cyclic: it is from PSII to PSI to NADP. There is also a cyclic flow of electrons that harnesses more energy but does not yield reduced NADP (see Figure 35).

Figure 35 Cyclic photophosphorylation

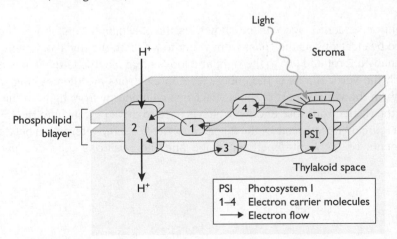

OCR A2 Biology

Table 6 Comparison of cyclic and non-cyclic photophosphorylation

Feature	Cyclic photophosphorylation	Non-cyclic photophosphorylation
Photosystem	I	I and II
Photolysis	No	Yes
Electron donor	P700 in photosystem I	Water
Last electron acceptor	P700 in photosystem I	NADP
Products	ATP	ATP; reduced NADP; oxygen

The light-independent stage

The light-independent stage occurs in the stroma and involves the fixation of carbon dioxide. There are animations of this cycle of reactions at, for example:

www.johnkyrk.com/photosynthesisdark.html

http://highered.mcgraw-hill.com/sites/0070960526/student_view0/chapter5/animation_quiz_1.html

Carbon dioxide diffuses into leaves through stomata, dissolves in water in the cell wall and diffuses into cells and into chloroplasts. The enzyme in the stroma that fixes carbon dioxide is **ribulose bisphosphate carboxylase oxygenase**, usually abbreviated to **rubisco**. This enzyme combines carbon dioxide with the 5-carbon acceptor compound **ribulose bisphosphate** (**RuBP**) to form a 6-carbon compound that immediately breaks down to form two molecules of the 3-carbon compound **glycerate 3-phosphate** (**GP**). These molecules are reduced to form **triose phosphate** (**TP**) which can be converted into a range of compounds including more RuBP, the carbon dioxide acceptor molecule. Since triose phosphate can be used to regenerate ribulose bisphosphate the reactions form a cycle. The cycle is named after one of the American scientists who discovered it — Melvin Calvin.

Figure 36 shows the Calvin cycle. The three main processes are as follows:

- **carboxylation** — combination of carbon dioxide with RuBP to form GP:

  CO_2 + RuBP (5C) $\rightarrow$ 2 × GP (3C)

- **reduction** of GP to TP using reduced NADP from the light-dependent stage
- **regeneration** by **synthesising** the carbon dioxide acceptor, RuBP

Energy in the form of ATP produced by the light-dependent stage drives the Calvin cycle. GP is phosphorylated by ATP to give TP and ribulose phosphate (RP) is phosphorylated to regenerate RuBP.

TP molecules are exported from chloroplasts into the cytosol by an antiport carrier protein in exchange for phosphate ions. TP molecules may be combined to form hexose phosphates, which are used to synthesise sucrose for transport, starch for storage and cellulose for cell walls. They are also converted into organic acids that combine with ammonia to give amino acids. Fatty acids are also made and converted into triglycerides for storage and phospholipids for membranes.

Examiner tip

Remember that carbon dioxide is one of the raw materials of photosynthesis.

Examiner tip

You can use the abbreviation 'rubisco' when writing about the enzyme that catalyses the fixation of carbon dioxide. The enzyme accepts carbon dioxide and oxygen in its active site, hence the inclusion of carboxylase and oxygenase in its name. You only need to know about the carboxylase function.

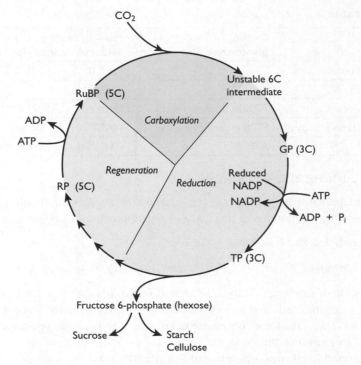

Figure 36 The Calvin cycle

Synoptic links

Examiners like to test your knowledge of the Calvin cycle by setting questions on graphs that show changes in the concentrations of RuBP, GP and TP when conditions such as light intensity, carbon dioxide concentration and temperature change. See pp. 90–91 for information on these graphs.

The light-independent stage is sometimes called the 'dark stage' or 'dark reaction'. This does not mean that it occurs in plants when it is dark. As light intensity decreases in the evening, there is less energy available from the light-dependent stage and the reactions of the Calvin cycle slow down and then stop.

The rate of photosynthesis is dependent on temperature. In Unit F212 you studied temperature as one of the factors that influence enzyme action. There are many enzymes involved in the Calvin cycle so the rate of the light-independent stage is temperature dependent. See p. 59 for a description of the effect of temperature as a limiting factor of photosynthesis.

Focus on practical skills: investigating the rate of photosynthesis

There are various ways in which the rate of photosynthesis can be investigated:
- absorption of carbon dioxide
- production of oxygen
- increase in dry mass

The easiest method is the production of oxygen. In schools and colleges, aquatic plants such as *Cabomba* spp. and Canadian pondweed, *Elodea canadensis*, are used. The apparatus used to measure the rate of photosynthesis is shown in Figure 37.

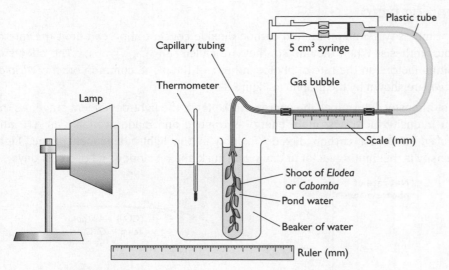

Figure 37 A photosynthometer: the light intensity is varied by moving the lamp to different distances from the plant

This apparatus can be used to investigate the effect of three factors on photosynthesis:
- light intensity
- carbon dioxide concentration
- temperature

In setting up the apparatus, the following precautions must be taken:
- The water should be aerated so that it is saturated with oxygen. This makes sure that any oxygen produced by the plant does not dissolve, but comes out of solution to form bubbles that can be collected.
- The water must contain some sodium hydrogencarbonate to provide a supply of carbon dioxide.
- The apparatus must be left for a while after the conditions are changed to make sure that the rate of photosynthesis is constant for those conditions.
- When investigating change in one factor, the others must be kept constant.

The rate of photosynthesis is calculated by dividing the length of the gas bubble produced by the time taken to collect it. At least three readings should be taken for each set of conditions and the mean calculated.

The effect of variables on the rate of photosynthesis is investigated as follows:
- **Temperature.** The temperature of the water can be changed by placing the beaker in a thermostatically controlled water bath at different temperatures and checking the temperature of the water in the test tube.
- **Carbon dioxide concentration.** Different quantities of sodium hydrogen-carbonate are added to the water in the test tube.

Bubbles of gas rise from the cut end of the stem to collect in the capillary tubing. Pulling gently on the syringe plunger causes the gas bubble to move further along the tubing so its length can be measured against the scale.

Knowledge check 22
Explain how the volume of gas produced in the apparatus shown in Figure 37 is calculated.

- **Light intensity.** The lamp is positioned at different distances from the plant. The light intensity can be measured with a light meter or it can be estimated by calculating $1/d^2$ where d = the distance between plant and lamp.

Limiting factors

Light intensity, temperature and carbon dioxide concentration can limit the rate of photosynthesis. When they do so, they are called limiting factors. The effects of limiting factors on the rate of photosynthesis of the plant, common orache, *Atriplex patula*, are shown by the graphs in Figure 38.

Light intensity determines the energy available for the light-dependent stage. As the light intensity increases, more energy is trapped and made available as ATP and reduced NADP for carbon dioxide fixation in the light-independent stage. Light intensity is the limiting factor at dawn and dusk and on cloudy and overcast days.

A limiting factor is any environmental factor that restricts a process, such as photosynthesis or growth. When all other factors are favourable, the factor at or closest to its minimum is the factor that is limiting. On hot, sunny days carbon dioxide concentration is the limiting factor for photosynthesis.

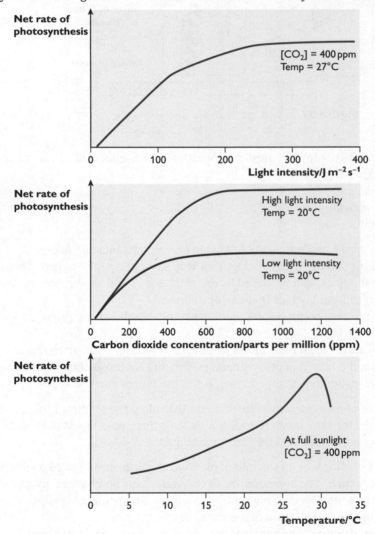

Figure 38 The effects of light intensity, carbon dioxide concentration and temperature on the rate of photosynthesis of common orache, *Atriplex patula*

Carbon dioxide concentration limits the rate of photosynthesis because it is the raw material for the light-independent stage. If there is a limited supply of carbon dioxide, small quantities of triose phosphate are produced and there is a limited demand for ATP. Therefore, the light-dependent stage is slow and little oxygen is produced. The carbon dioxide concentration of the air is 0.04% or 400 parts per million. On warm sunny days, or deep inside the canopy of a forest, or in a crop where competition for the gas is intense, it is likely that carbon dioxide concentration decreases and so it becomes the limiting factor.

The reactions of the light-independent stage are catalysed by enzymes. At low temperatures, the rate of these reactions is slow. As the temperature increases, the reaction rate increases. At high temperatures, enzymes are denatured. At low temperatures, ice crystals form within cells, break cell membranes and so destroy cells.

You can expect to carry out quantitative experiments on photosynthesis in Unit F216. You may need this information on limiting factors in order to analyse and interpret your results in the quantitative and evaluative tasks in Unit F216. Examples 19, 20 and 21 in the Unit guide to Unit F213/F216 are based on different ways of investigating photosynthesis.

Examiner tip

Note the shape of the lines on the first two graphs: a slope and a plateau. These are explained in terms of limiting factors. In the slope part, the rate is limited by the factor that is the independent variable. In the plateau, the rate is limited by another factor.

Knowledge check 24

Many crops are grown in greenhouses (glasshouses). The environmental conditions can be controlled to maximise production. Suggest how *three* such factors may be controlled to maximise rates of photosynthesis.

Summary

- Autotrophic organisms fix carbon dioxide using energy from the Sun or from simple chemical reactions. Plants use light energy to produce complex organic molecules in photosynthesis. Heterotrophs take in fixed carbon in the form of organic molecules.

- Photosynthesis provides triose phosphate which is used to make organic compounds in plants, including carbohydrates and fats that are respired to provide ATP.

- Photosynthesis is a two-stage process that occurs entirely within chloroplasts.

- The light-dependent stage occurs in the grana. Photosynthetic pigments absorb light energy and transfer it to chlorophyll *a* molecules at the centre of photosystems I and II (PSI and PSII). Electrons emitted by these molecules travel through the electron transport chain with pumping of protons into thylakoids and the formation of reduced NADP. These protons diffuse through ATP synthase which produces ATP. In cyclic photophosphorylation, electrons flow from and return to PSII with the formation of ATP; in non-cyclic photophosphorylation, electrons from the photolysis of water reduce NADP with the generation of ATP.

- Oxygen is a by-product of photolysis and is used in aerobic respiration; the excess diffuses to the surroundings.

- The light-independent stage occurs in the stroma. ATP and reduced NADP drive the reactions of the Calvin cycle. The enzyme rubisco catalyses the fixation of carbon dioxide as glycerate 3-phosphate (GP). Other steps in the cycle are triose phosphate (TP) formation and regeneration of the carbon dioxide acceptor RuBP.

- Triose phosphate is used to make carbohydrates (e.g. sucrose, starch and cellulose), lipids and amino acids. Most is recycled to RuBP.

- Limiting factors for the rate of photosynthesis are light intensity, carbon dioxide concentration and temperature. Changes in all three affect the concentrations of GP, RuBP and TP in the stroma.

- The effects of limiting factors on the rate of photosynthesis are investigated by determining the rate of oxygen production by measuring the volume of gas collected from aquatic plants. Most of this gas is oxygen.

Module 4: Respiration

Key concepts you must understand

Plants, animals and microorganisms respire in order to release energy from carbon compounds and make it available for a variety of functions:

- biosynthesis, e.g. protein, carbohydrate, fat and nucleic acid synthesis
- active transport, e.g. sodium–potassium ion pumps
- movement, e.g. muscle contraction to move part of the body or the whole body, movement of chromosomes during mitosis and meiosis, movement of phagocytic vacuoles and secretory vesicles; movement of flagella (prokaryotes), movement of cilia and undulipodia (eukaryotes)
- maintenance of body temperature in endotherms
- raising the energy level of glucose at the start of glycolysis (see p. 63)

Energy is available in compounds, such as carbohydrates, fats and proteins, which on oxidation release energy. When you burn foods in oxygen to find out how much energy they provide (see p. 70), all the energy is released at once. This generates so much heat that, if it happened in cells, proteins would be denatured. Enzymes catalyse reactions in which small changes occur. To release all the energy from a compound such as glucose requires many enzyme-catalysed reactions. Some of these transfer energy directly to ATP. In some, energy is transferred indirectly in the reduction of the coenzymes NAD and FAD.

Animations of the stages of respiration can be found at:

www.johnkyrk.com

http://highered.mcgraw-hill.com/sites/0070960526/student_view0/chapter5/animations.html

Key facts you must know

ATP

Adenosine triphosphate (ATP) is the universal energy currency. It is a compound that transfers energy within cells. We have already seen that it is involved in photosynthesis.

The human body has about 75 g of ATP at any one time. It is not stored and is not transported between cells; it is made as and when it is required. The molecules are small and water soluble, so ATP moves around easily within cells. It is a molecule that many enzymes 'recognise' as it fits into active sites and acts as a coenzyme in many reactions by transferring phosphate groups. The turnover of ATP each day is huge as each molecule is used and recycled continually. You will come across textbooks and websites that refer to ATP as a 'high-energy compound'. It is no such thing. As a molecule it releases less energy when hydrolysed than many other biochemical molecules. You will also see 'ATP has high-energy bonds'. This is not true either. When ATP transfers a phosphate group the energy transferred comes from the whole molecule not just from the bond with the terminal phosphate.

Examiner tip

Respiration is a chemical process catalysed by enzymes that transfers energy from carbon compounds to ATP. It occurs in all living cells. Do not confuse respiration with breathing and gaseous exchange, which are physical processes.

Knowledge check 25

Distinguish between respiration and gaseous exchange in a mammal.

When ATP is hydrolysed, the terminal bond is broken with the transfer of energy:

$$ATP + H_2O \rightarrow ADP + P_i + 30.5 \, kJ \, mol^{-1}$$

This reaction is always coupled with another reaction so that energy is transferred. ATP is good at transferring phosphate groups and transferring energy, which is why it is often called the 'energy currency' of cells.

The structure of ATP is shown in Figure 39.

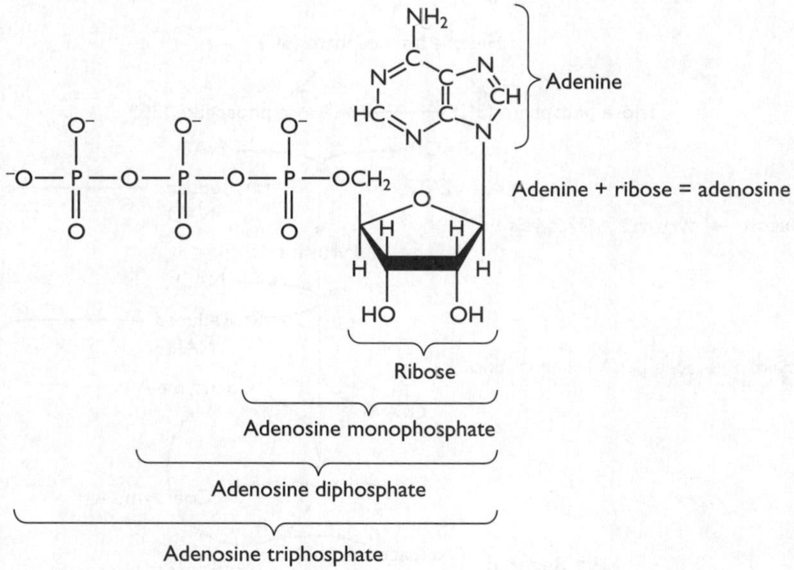

Figure 39 ATP is a triphosphate nucleoside composed of the base adenine, the pentose sugar ribose and three phosphate groups

Knowledge check 26

Explain why ATP is described as the 'universal energy currency'.

Table 7 Roles of ATP

Role	Detail of process	Example
Active transport	ATP provides phosphate that binds to carrier proteins so that they change shape to transport substances against their concentration gradients	Sodium–potassium ion pump
Producing reactive compounds	ATP transfers phosphate to an unreactive compound	Phosphorylation of glucose in glycolysis
	ATP transfers AMP to an unreactive compound	Biosynthesis: activation of amino acids so they attach to transfer RNA in protein synthesis
Binding to a protein	ATP activates a protein	ATP-gated potassium channel protein (see p. 33)
	ATP inhibits a protein	A key enzyme of respiration is inhibited by ATP
Providing energy for movement	ATP binds to proteins causing them to change shape and bring about movement	All movement within cells, e.g. cilia; undulipodia; muscle contraction and movement of vesicles

We have already seen that ATP is produced in photophosphorylation. During respiration, ATP is produced in two processes:

- substrate-level phosphorylation
- oxidative phosphorylation

Knowledge check 27

State the precise locations in a plant cell where ATP is made.

Aerobic respiration

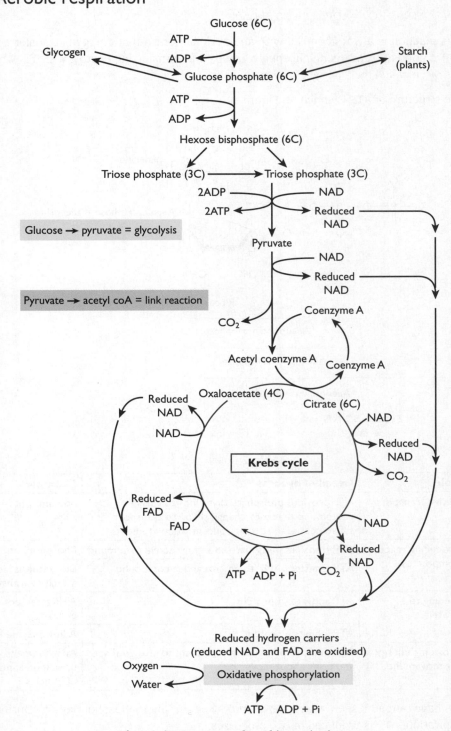

Figure 40 Summary of aerobic respiration

Respiration involves the transfer of energy from organic molecules, such as carbohydrates, fats and proteins, to ATP.

Respiration depends on coenzymes. Coenzyme A takes part in a number of reactions and passes 2-carbon fragments into the Krebs cycle. During respiration, there are a number of dehydrogenation reactions in which hydrogen atoms are released from compounds. These reduce the coenzyme NAD, which is mobile and transfers hydrogen atoms to the internal membranes of the mitochondria. NAD is a hydrogen carrier.

Aerobic respiration of glucose occurs in four stages in cells:
(1) Glycolysis — in the cytosol
(2) Link reaction — in the matrix of mitochondria
(3) Krebs cycle — in the matrix of mitochondria
(4) Oxidative phosphorylation — across the inner membrane of mitochondria

Aerobic respiration is summarised in Figure 40.

In the first part of glycolysis, glucose is converted into hexose (fructose) bisphosphate (a 6-carbon compound) which is broken down into two molecules of triose phosphate (a 3-carbon compound). The reactions that follow occur *twice* for each molecule of glucose. Remember this when studying the yield of ATP on p. 68.

Glycolysis

Glycolysis literally means the 'splitting of glucose'. The first reaction involves phosphorylating glucose to give glucose phosphate, which is then phosphorylated to hexose (fructose) bisphosphate. (Respiration may begin with glycogen or starch, in which case phosphorylase enzymes first break down these molecules to glucose phosphate.) The next step is the conversion of each hexose bisphosphate molecule into two molecules of triose phosphate, which are then oxidised to pyruvate. During these steps a dehydrogenation reaction occurs in which NAD is reduced. Energy from glucose is transferred to NAD and is later transferred to ATP. In the last steps of glycolysis, ATP is synthesised by substrate-level phosphorylation.

The products per molecule of glucose are:
- 2 × reduced NAD
- 4 × ATP (two ATP were used at the start, so the net gain is 2 ATP)
- 2 × pyruvate

The end product of glycolysis is pyruvate. This is energy rich and is respired in mitochondria if oxygen is present. Pyruvate molecules enter mitochondria through symport channel proteins in the inner membrane. This uptake is driven by the proton gradient across this membrane (see p. 66). Pyruvate molecules enter a reaction that links glycolysis to the Krebs cycle.

Figure 41 shows the structure of a mitochondrion and the exchanges with the cytosol. The outer membrane of mitochondria is freely permeable to many substances. The inner membrane is much more selective, controlling what passes in and out.

Examiner tip
To help your revision, make a large outline diagram of respiration. As you read this guide and other books and websites, add more details. Do not worry about names of all the intermediate compounds you will find, concentrate on the principles that are given here.

Examiner tip
Note the use of ATP at the start of glycolysis. You may be asked why the *net* production of ATP in glycolysis is only two molecules.

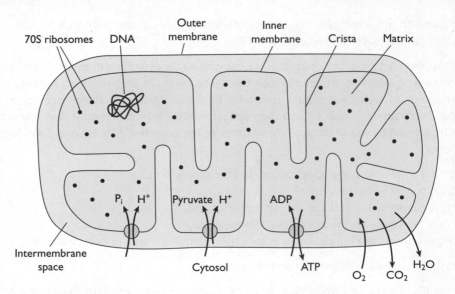

Figure 41 The structure of a mitochondrion and the exchanges with the cytosol

The link reaction

Pyruvate is a 3-carbon compound. During the link reaction (see Figure 40), it is converted into the 2-carbon acetyl (ethanoyl) group (sometimes known as 'active acetate'), which combines with coenzyme A to make acetyl coenzyme A. This conversion involves decarboxylation (removal of carbon dioxide) and dehydrogenation (removal of hydrogen, i.e. oxidation). The hydrogen atoms are accepted by NAD, and reduced NAD is formed.

Pyruvate dehydrogenase is a complex made of a number of polypeptides for the three intermediate reactions that form the link between glycolysis and the Krebs cycle. Acetyl coenzyme A transfers the 2-carbon fragment (acetyl group) into the next stage, which is the Krebs cycle (named after Sir Hans Krebs, who discovered it).

The products of the link reaction per molecule of glucose are:
- 2 × reduced NAD
- 2 × CO_2
- 2 × acetyl (2C) coenzyme A

Krebs cycle

Acetate is passed from coenzyme A to oxaloacetate to form citrate. Enzymes in the mitochondrial matrix catalyse reactions to regenerate oxaloacetate which is the acceptor substance for acetate. During the Krebs cycle, decarboxylation and dehydrogenation reactions occur. Carbon dioxide diffuses out of mitochondria; hydrogen atoms reduce the coenzymes NAD and FAD.

Following the Krebs cycle round (Figure 40), the processes are as follows:
- reaction between acetyl coenzyme A and oxaloacetate — coenzyme A delivers a 2C fragment (acetyl group) into the cycle to form citrate (6C)
- decarboxylation (× 2) — removal of carbon dioxide
- dehydrogenation (× 4):

Examiner tip

The enzyme complex that catalyses the link reaction is known as pyruvate dehydrogenase. Although it decarboxylates pyruvate, it is not known as a decarboxylase. Pyruvate decarboxylase is a different enzyme that catalyses a different reaction.

- removal of hydrogen from intermediate substances, which are oxidised
- reduction of NAD (× 3)
- reduction of FAD (× 1)
- ATP synthesis (× 1) — substrate-level phosphorylation
- regeneration of oxaloacetate (4C compound)

Products per molecule of glucose (remember, there are two 'turns' of the cycle per molecule of glucose):
- 4 × carbon dioxide
- 2 × reduced FAD
- 6 × reduced NAD
- 2 × ATP

Oxidative phosphorylation

Glycolysis, the link reaction and Krebs cycle generate reduced NAD and reduced FAD. There is only a small quantity of these coenzymes in a cell, so in order for the reactions to continue it is necessary for these reduced coenzymes to be oxidised. This oxidation transfers energy to ATP and occurs during oxidative phosphorylation, which is the fourth stage of aerobic respiration (see Figure 42).

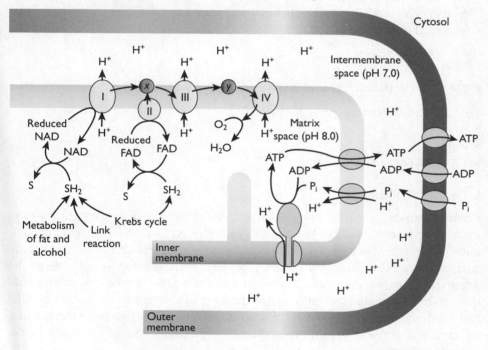

SH_2 = intermediate substances in the link reaction, Krebs cycle and metabolism of fat and alcohol. These substances are dehydrogenated to form reduced hydrogen carriers (reduced FAD and reduced NAD), which are oxidised by complexes I and II

I, II, III and IV are protein complexes
x and y are electron carrier molecules

Complexes I, III and IV pump protons; complex II passes electrons from reduced FAD to complex III via electron carrier x. Complex II is not a proton pump

Figure 42 Oxidative phosphorylation

Unit F214: Communication, Homeostasis and Energy

Knowledge check 28

Succinate dehydrogenase catalyses the dehydrogenation of succinate in the Krebs cycle. Malonate has a molecular structure similar to succinate. Explain why the rate of the reaction slows when malonate is added to a reaction mixture of succinate and the enzyme.

Reduced coenzymes provide hydrogen atoms. Reduced NAD and FAD are oxidised to provide protons and electrons. The electrons pass along the electron transport chain. The energy released is used to pump protons from the mitochondrial matrix into the intermembrane space. This creates a proton gradient which is harnessed to make ATP. The protons diffuse from the intermembrane space to the matrix through ATP synthase which generates ATP. The energy that drives the whole process is the energy from the respiratory substrates that were oxidised during the previous three stages.

Electrons combine with oxygen and protons to form water. This reaction is catalysed by cytochrome oxidase. Oxygen is the final electron acceptor. The water produced in the reaction is sometimes known as metabolic water.

The processes in oxidative phosphorylation are as follows:
- Oxidation of coenzymes NAD and FAD — coenzymes are recycled for use in metabolic pathways in the matrix of the mitochondrion: link reaction, Krebs cycle and the metabolism of fats and alcohol.
- Electrons flow along the electron transport chain — this provides energy for the active transport of protons.
- Protons are pumped from the matrix into the intermembrane space through three protein complexes (labelled I, III and IV on Figure 42).
- Protons diffuse through ATP synthase — ATP synthase catalyses ATP synthesis:

$$ADP + P_i \rightarrow ATP$$

- The final electron acceptor is oxygen, which is reduced to water.

The products of oxidative phosphorylation are:
- ATP
- water
- NAD and FAD (oxidised forms)

Chemiosmosis

Chemiosmosis is the process by which protons are pumped to create a gradient and the energy in that gradient is used to synthesise ATP. It occurs in chloroplasts, mitochondria and bacteria. Protons are pumped through protein complexes that are part of the electron transport chain. The high concentration of protons in the intermembrane space in mitochondria and in the thylakoid space in chloroplasts gives rise to a potential difference across the inner mitochondrial membrane and the thylakoid membrane. This gives an electrochemical gradient across these membranes, which are impermeable to protons except by diffusion through channels in the enzyme ATP synthase.

The proton gradient and membrane potential are a **proton-motive force** that drives ATP synthesis. This electrochemical gradient acts like a battery as a store of potential energy — in this case, for the synthesis of ATP. In mitochondria, it is recharged continually using energy from oxidised food. In chloroplasts, it is recharged during the day using light energy absorbed by photosynthetic pigments. It cannot function in chloroplasts in the dark.

Knowledge check 29

State the precise locations of the four stages of aerobic respiration.

Examiner tip

To see ATP synthase in action, see the animation at www.iubmb-nicholson.org/swf/ATPSynthase.swf

Knowledge check 30

State the roles of coenzymes in respiration.

Evidence for the process of chemiosmosis

Chemiosmosis was first proposed in the early 1960s as a mechanism for ATP synthesis. There are several lines of evidence for this mechanism:

- The pH in the intermembrane spaces is lower than in the mitochondrial matrix. In chloroplasts the pH in the thylakoid spaces is lower than in the stroma.
- Isolated chloroplasts suspended in a sucrose solution and illuminated turn the pH of the solution alkaline as protons are removed from the medium and pumped into the thylakoids.
- Given ADP and phosphate, isolated grana kept in an acid medium can make ATP when transferred to an alkaline solution in the dark (see Question 5 on p. 89 where this is explained in detail).
- Artificial membranes made from phospholipids, light-driven protein pumps from bacteria and ATP synthase from cardiac muscle mitochondria produce ATP when the membranes are exposed to light.

Anaerobic respiration

If oxygen is not available, then the last stage of oxidative phosphorylation cannot occur because it needs oxygen as the final oxygen acceptor. Reduced NAD and FAD are not oxidised and the Krebs cycle and the link reaction stop. Pyruvate is no longer moved into the mitochondria. Pyruvate could be excreted, but there would have to be a way to recycle the NAD that takes part in glycolysis. Respiration continues in the absence of oxygen to recycle NAD, using pyruvate. Figure 43 shows what happens to the TP and pyruvate produced in glycolysis in anaerobic respiration in mammals and yeast. Remember that in both cases 2 molecules of ATP were used in earlier reactions (see Figure 40).

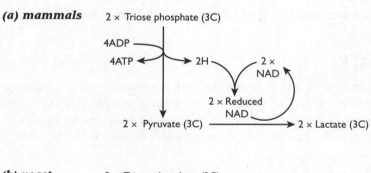

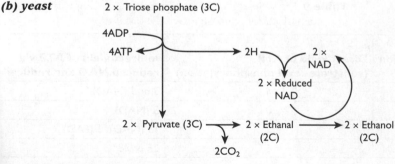

Figure 43 (a) Fate of pyruvate in anaerobic respiration in mammals
(b) Fate of pyruvate in anaerobic respiration in yeast

Table 8 Comparison between anaerobic respiration in mammals and yeast

Feature	Mammal	Yeast
Decarboxylation (to give carbon dioxide)	No	Yes
Oxidation of reduced NAD	Yes	Yes
Hydrogen acceptor	Pyruvate (3C)	Ethanal (2C)
Products per molecule of glucose	2 × lactate (3C)	2 × ethanol (2C); 2 × CO_2
Net gain of ATP per molecule of glucose	2	2
Reversible reaction(s)	Yes (in liver and heart muscle)	No

The overall equation for anaerobic respiration in mammals is:

$$C_6H_{12}O_6 \rightarrow 2CH_3CHOHCOOH$$

with a net gain of 2ATP per molecule of glucose.

This occurs in muscle and certain other tissues when there is a shortage of oxygen. The advantage of anaerobic respiration in muscle is that energy is provided even though there is limited availability of oxygen. The lactate produced diffuses from muscle tissue into the blood. Cardiac muscle in the heart uses lactate in respiration. The rest is recycled to glucose and glycogen in the liver, so the energy is not lost. Anaerobic respiration in mammals is, therefore, reversible.

In yeasts and plants, pyruvate is first decarboxylated by the enzyme pyruvate decarboxylase to form ethanal which acts as a hydrogen acceptor to form ethanol. The overall equation for anaerobic respiration in yeast is:

$$C_6H_{12}O_6 \rightarrow 2C_2H_5OH + 2CO_2$$

with a net gain of 2ATP per molecule of glucose.

Yeast and plants cannot metabolise ethanol so this form of anaerobic respiration is irreversible.

Yield of ATP

Table 9 shows how to calculate the theoretical maximum yield of energy in the form of ATP from aerobic respiration. It is currently thought that during oxidative phosphorylation each reduced NAD molecule provides the energy to make 2.5 ATP molecules and each reduced FAD molecule provides the energy to make 1.5 ATP molecules.

Table 9 The yield of ATP from the complete oxidation of a molecule of glucose in aerobic respiration

Stage of aerobic respiration	Input of ATP (phosphorylation of hexose)	Direct yield of ATP (substrate-level phosphorylation)	Indirect yield of ATP via reduced NAD and reduced FAD
Glycolysis	−2	4	3 or 5 (NAD)*
Link reaction		0	5 (NAD)
Krebs cycle		2	15 (NAD); 3 (FAD)
Totals	−2	6	26 or 28*

This gives a theoretical maximum yield of 32 − 2 = 30 molecules of ATP. (*or 32 depending on how the hydrogens from reduced NAD from glycolysis enter the mitochondria. There are two methods for this, which give rise either to 3 or to 5 molecules of ATP.)

However, this theoretical maximum is rarely, if ever, reached because:

- some intermediates in the metabolism of glucose are converted into other substances rather than being broken down completely
- glucose is not metabolised in isolation, since through the 'metabolic funnel' fatty acids and amino acids are also metabolised
- the proton gradient is used to power the movement of substances into the matrix, e.g. pyruvate and phosphate ions
- chemiosmosis is not efficient, as some protons are lost through the outer mitochondrial membrane

The important point to remember here is not the calculation in Table 9, but the difference between the net yield in aerobic respiration (about 30 ATP molecules) and the net yield in anaerobic respiration, which is 2 ATP molecules. This is because none of the energy transfer reactions of the link reaction, Krebs cycle and oxidative phosphorylation occur in anaerobic respiration.

The transfer of energy to ATP is not 100% efficient — much is 'lost' as heat. Endotherms can retain this heat in the body to maintain a constant body temperature; ectotherms lose the heat to their surroundings.

Metabolic funnel

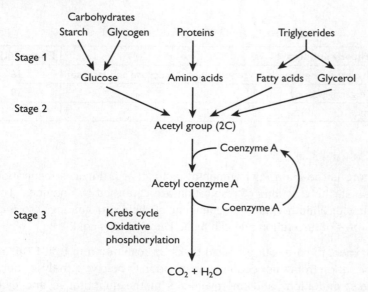

Figure 44 The 'metabolic funnel': the respiratory substrates are converted into acetyl coenzyme A which then enters the Krebs cycle

At A-level we deal with the respiration of glucose, but animals respire glycogen, lots of fat and, in the case of carnivores, quite a lot of protein. These substances are known as **respiratory substrates**. These compounds have to be converted into a form that can be respired. Figure 44 shows the 'metabolic funnel', which is an outline of the various pathways that lead to the Krebs cycle and oxidative phosphorylation, i.e. to the pathways that are common to the respiration of all respiratory substrates.

The calculations for ATP yield in respiration in Table 9 are for glucose as the respiratory substrate. Other carbohydrates go through stage 1 of the metabolic funnel to form glucose. The reactions in stage 1 do not release any energy, so the energy yield per gram is the same for all carbohydrates.

Some cells, such as brain cells and red blood cells, respire only glucose. Others, such as cardiac muscle, respire fats. Fats are energy rich. The molecules are composed mainly of carbon and hydrogen so there are many C–H bonds. When these are oxidised, there is the release of much energy per gram. The breakdown of a triglyceride in respiration involves breaking the ester bonds between glycerol and the three fatty acids and then removing 2-carbon fragments from fatty acids as 'active acetate'. These are then sent to the Krebs cycle via coenzyme A. This preliminary stage is known as β-oxidation; it occurs in mitochondria and involves the transfer of energy to reduced NAD.

Proteins are broken down into amino acids in stage 1. The amino acids must first be deaminated (see p. 40) before they can be respired. An amino acid that has had its amino group removed is an organic acid. Many of these organic acids are similar to those in the Krebs cycle. Therefore, respiring amino acids produces almost as much energy as respiring carbohydrates.

Table 10 shows the energy values in $kJ\,g^{-1}$ for three respiratory substrates. The energy values of substrates are found by calorimetry — burning substances in oxygen and recording the heat released. The energy content depends on the number of hydrogen atoms per molecule that are available to reduce NAD and FAD. Lipids (fats) have the most.

Table 10 Energy values of respiratory substrates

Respiratory substrate	Energy/$kJ\,g^{-1}$
Carbohydrates, e.g. starch, glycogen, glucose, sucrose and lactose	16
Lipids, e.g. triglycerides	39
Proteins	17

Knowledge check 32

Explain why the energy value of lipids is far greater than that for carbohydrates.

Synoptic links

Temperature is the main environmental factor that influences respiration. This is because all the reactions of respiration are catalysed by enzymes. The body temperature of endotherms is independent of their surroundings, so they have constant rates of respiration and can be active when it is cold.

You can expect to carry out practical tasks on respiration in Unit F216. You will need information from this section on respiration to help you analyse your results and suggest limitations and improvements in the quantitative and evaluative tasks.

Examples 22–26 and the A2 Quantitative and Evaluative tasks in the Unit guide for Units F213/F216 are about respiration. They should show you how to approach the practical tasks in Unit F216.

OCR A2 Biology

- All organisms respire, transferring energy from organic molecules such as glucose and fat to ATP, which provides energy for active transport, biosynthesis and movement.

- The four stages of aerobic respiration are: glycolysis (cytosol), link reaction, Krebs cycle (matrix of mitochondrion) and oxidative phosphorylation (cristae).

- Glycolysis begins with the phosphorylation of glucose by 2 × ATP to form hexose bisphosphate. This splits into two molecules of triose phosphate, which are oxidised to pyruvate with the production of 2 × reduced NAD; 4 × ATP are formed by substrate-level phosphorylation.

- In the link reaction, pyruvate is dehydrogenated to form reduced NAD and decarboxylated to form carbon dioxide and an acetyl group (2C).

- Coenzyme A transfers an acetyl group to oxaloacetate (4C) to form citrate (6C). In the Krebs cycle, the 2C compound is decarboxylated to form $2 \times CO_2$, oxidised to form 3 × NAD and 1 × FAD and oxaloacetate is regenerated. Substrate-level phosphorylation occurs to produce ATP. The link reaction and Krebs cycle occur twice for each glucose molecule.

- Reduced NAD and FAD are recycled by oxidative phosphorylation. Electrons flow along an electron transport chain with transfer of energy to pump protons from the matrix into the intermembrane space. Protons flow down their gradient through ATP synthase, which phosphorylates ADP to ATP. The final electron acceptor is oxygen with the formation of water.

- Chemiosmosis is the use of proton pumps to produce a proton gradient across membranes. The gradient drives the synthesis of ATP by ATP synthase. The many internal thylakoids in chloroplasts and folded cristae in mitochondria provide large surface areas for many protein complexes.

- Evidence for chemiosmosis includes differences in pH across membranes in chloroplasts and mitochondria as a result of proton pumping.

- Yields of ATP rarely match those expected because respiratory substrates, such as glucose and fat, are not always oxidised completely; energy is used in exchanging substances (e.g. pyruvate) between mitochondria and cytosol; some protons 'leak' through the outer mitochondrial membrane.

- In anaerobic respiration, pyruvate does not enter mitochondria so much less energy is transferred. In mammals, pyruvate is the final electron and hydrogen acceptor to recycle NAD and form lactate. In yeast and plants, pyruvate is decarboxylated to ethanal with the formation of carbon dioxide. Ethanal is the hydrogen acceptor and forms ethanol when NAD is recycled. The only ATP produced is during glycolysis, which explains the much lower yield than that in aerobic respiration.

- Lipids are the most energy-rich respiratory substrate as they have a high ratio of hydrogen to carbon and are more highly reduced than carbohydrates and proteins. On oxidation they transfer most energy per gram to ATP.

Questions & Answers

The unit test

The examination paper will be printed in a booklet, in which you will write all your answers. The paper will have about six questions, each divided into parts. These parts comprise several short-answer questions (no more than 4 or 5 marks each) and two questions requiring extended answers, for around 5 or 6 marks each. The unit test offers a total of 60 marks and lasts 1 hour and 15 minutes.

You can expect questions to cover more than one module of the unit, as in Question 5, which examines some of the content of Modules 3 and 4.

As you read through this section, you will discover that student A gains full marks for all the questions. This is so that you can see what high-grade answers look like. Notice how student A uses technical terms from the specification to good effect. Remember that the minimum for grade A is about 80% of the maximum mark (in this case around 48 marks). Student B makes a lot of mistakes — often these are ones that examiners encounter frequently. I will tell you how many marks student B gets for each question. If the overall mark for the paper is about 40% of the total (around 24 marks), the student will have passed at grade-E standard. Use these benchmarks when trying the questions yourself.

The quality of your written communication is assessed in two questions, which are indicated on the exam paper. You are expected to use technical terms correctly in the longer-answer questions.

On p. 95 there is a summary of the mistakes made by student B. This should help you to identify what the candidate should have done during revision and in the examination in order to gain a better mark.

Examiner's comments

Examiner comments on the questions are preceded by the icon ⓔ. They offer tips on what you need to do in order to gain full marks. Candidates' answers are followed by examiner's comments. These are preceded by the icon ⓔ and indicate where credit is due. In Student B's weaker answers they also point out areas for improvement, specific problems and common errors, such as lack of clarity, weak or non-existent development, irrelevance, misspellings, misinterpretation of the question and mistaken meanings of terms.

Question 1 **Thermoregulation**

The hypothalamus is the body's thermoregulation centre. It integrates information from central and peripheral thermoreceptors and controls the core body temperature. The set point for core body temperature can change.

The influence of peripheral thermoreceptors in the skin on the set points for heat loss by sweating and heat production by shivering was investigated. The body's core temperature was changed and the rate of sweating and the rate of heat production were measured at different skin temperatures. The results are shown in Figure 1.

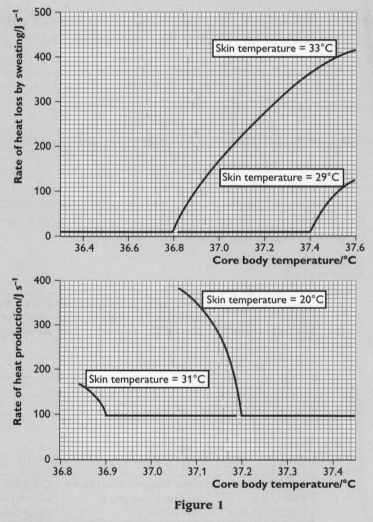

Figure 1

(a) (i)   Complete the table by showing the core body temperatures at which:
  – sweating begins when the skin temperatures are 33°C and 29°C
  – heat production by shivering begins when the skin temperatures are 31°C and 20°C.
  (Do not write anything in the shaded boxes in the table.)

  (2 marks)

Skin temperature/°C	Core body temperature at which sweating starts/°C	Core body temperature at which heat production by shivering starts/°C
33		
31		
29		
20		

ⓔ Use a ruler when interpreting graphs and when you are taking data from graphs, as in this question. Data quotes must be accurate.

(ii) Explain how the data shown in Figure 1 provide evidence for the idea that peripheral thermoreceptors act to reset the set point for core body temperature. (3 marks)

ⓔ Questions such as this that refer to a figure should always be answered with information taken directly from the figure. Note you are asked to provide *evidence*. You can describe the data qualitatively and/or you can use figures taken from the graph.

(b) Explain how temperature control in humans is an example of negative feedback. (3 marks)

Total: 8 marks

ⓔ Before answering this think about the vocabulary required for answers on homeostasis. Look back to the first section in the Content Guidance to find the appropriate words to use. Plan each of your answers, at least by noting the technical terms to use.

Student A

(a) (i)

Skin temperature/°C	Core body temperature at which sweating starts/°C	Core body temperature at which heat production by shivering starts/°C
33	36.8	
31		36.9
29	37.4	
20		37.2

Student B

(a) (i)

Skin temperature/°C	Core body temperature at which sweating starts/°C	Core body temperature at which heat production by shivering starts/°C
33	36.8°C	
31		36.9°C
29	37.4°C	
20		37.2°C

ⓔ Both candidates have read the graphs correctly and completed the tables. Student B has put in the units (°C) which is not necessary, but is unlikely to be penalised. However, you should never include units in this way in a table. Units should appear only in the column and row headings. Student B gains 2 marks.

Student A

(a) (ii) Peripheral temperature receptors detect temperature changes in the skin. When the air temperature changes the skin either absorbs heat or loses it to the surroundings. When the skin temperature is above 30°C, heat production by shivering does not start until the core body temperature reaches a low of 36.9°C, but at 20°C it starts at the higher temperature of 37.2°C. The same applies to sweating: when the skin temperature is above 30°C sweating starts at the lower core temperature of 36.8°C as opposed to 37.4°C when the skin temperature is less than 30°C. This shows that the set point is lower (36.8°C to 36.9°C) when the surroundings are hot and 37.2°C to 37.4°C when it is colder.

Student B

(a) (ii) As the body core temperature increases, sweating starts at 36.8°C when the skin temperature is above 30°C and at 37.4°C when 4°C lower. As the core body temperature increases, heat production by shivering stops at 36.9°C when the skin is at 31°C, while it does not stop until 37.2°C when it is 20°C.

ⓔ Adjusting the set point is a challenging concept. A question this difficult would not be at the beginning of the unit test, it would be towards the end, but in this guide the questions follow the sequence of topics in the specification. Student A has made good use of the information in the question, using the term set point, realising that it is adjusted depending on the air temperature and using the figures from the table. It is a good idea to annotate graphs carefully before answering the question. If you do this, then your understanding of graphs should be much better. If you look at Figure 1, you should be able to see the temperature ranges that student A describes. Student B has given appropriate data, but has not answered the question. You may be asked to explain how data support or refute a theory, statement, hypothesis or prediction and this involves careful analysis and interpretation of the data achieved by looking for trends and patterns in graphs and tables. Ruling some lines at temperatures at which the two processes start would be helpful — you probably need to do that to take accurate readings for (a)(i) anyway. Student B has made use of the data, for 1 mark.

Student A

(b) Negative feedback is a process that works by counteracting any change in the conditions in the body so that those conditions remain fairly stable. Thermoregulation is an example of negative feedback because core body temperature is kept within narrow limits. There is a set point that is determined by the hypothalamus, which has central thermoreceptors to monitor the core body temperature and receives information from peripheral thermoreceptors in the skin. When the body temperature falls below the set point, the hypothalamus coordinates the conservation of heat by reducing the sweating. If the temperature decreases further or if the air is cold, the hypothalamus stimulates heat production by shivering in the muscles. These corrective actions help to maintain the body temperature. When the temperature rises above the set point, the hypothalamus stimulates heat loss by sweating. This reduces the body temperature so it is close to the set point. The diagram shows the main points about negative feedback in thermoregulation.

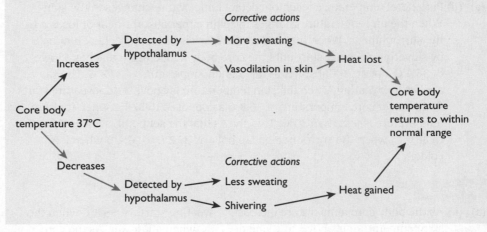

Student B

(b) When it is hot, the body loses heat. It does this by sweating. When the sweat is on the skin surface it absorbs heat and evaporates. This cools down the body so that the core temperature stays at 37°C. When it is cold, the body stops sweating so that heat is not lost from the body. Muscles contract in shivering to produce heat to help keep the body temperature at 37°C. Keeping the body temperature at 37°C is negative feedback.

ⓔ Student A has used the term negative feedback correctly. The inclusion of a diagram is a good idea since this is a difficult concept to explain. Student A has written a long answer that would probably need extra space. Examiners mark your scripts online. While marking questions, they see only the space where you should write your answer, but they will look at the whole page or at the additional page at the back of your script if you alert them to this. Student B has described how the body loses, conserves and produces heat, but has not explained how this is controlled. Negative feedback is not keeping the body temperature at 37°C — that is homeostasis. As the information in Figure 1 makes clear, the set point for body temperature is a range and the actual body temperature is maintained within narrow limits, but is not kept constant at 37°C. Student B does not gain any marks.

ⓔ **Student B gains 3 marks out of 8 for Question 1.**

Question 2 **Nerve transmission**

Figure 2 shows a motor neurone.

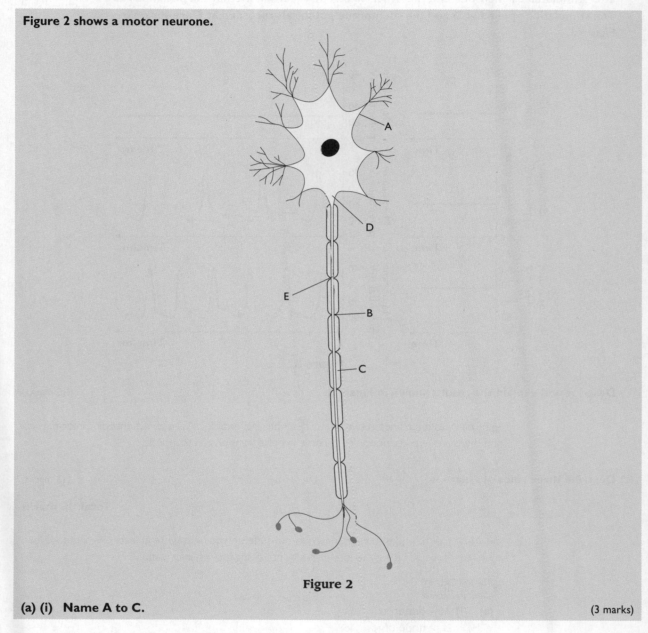

Figure 2

(a) (i) Name A to C.   (3 marks)

ⓔ 'Naming of parts' questions are common for this topic. These are easy marks, so make sure you remember the diagrams from your notes and write the correct names.

(ii) Explain how the resting potential is maintained across the membrane of a neurone.   (4 marks)

ⓔ It is probably best to start this answer by giving the value of the resting potential and then explain that it is due to the distribution of ions across the membrane. There are 4 marks — so make sure that you make four different points.

(b) Electrodes were placed at **D** and **E** on the motor neurone shown in **Figure 2**. The electrode at **D** stimulated the neurone and the other electrode recorded membrane potentials at **E**. The stimulation applied at **D** and the membrane potentials recorded at **E** are shown in **Figure 3**.

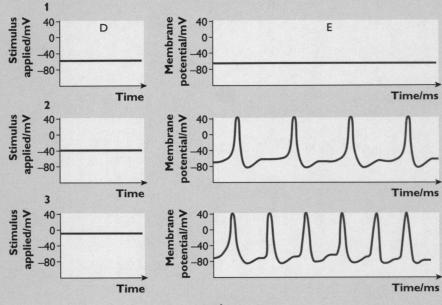

Figure 3

Describe and explain the results shown in Figure 3. (5 marks)

ⓔ This question comes in two parts: 'describe' and 'explain'. This is about encoding information and there are two key principles of nerve impulse transmission to include.

(c) Describe three roles of synapses. (3 marks)

Total: 15 marks

ⓔ Note that this question says 'describe' not state, so you *must* write at least one sentence for each role. It would be easy to lose marks by not giving enough information.

> **Student A**
>
> **(a) (i)** A — dendron
> B — node of Ranvier
> C — myelin

> **Student B**
>
> **(a) (i)** A — dendron
> B — axon
> C — Schwann cell

OCR A2 Biology

ⓔ Both candidates gain 3 marks. Label line B points to the axon at a node of Ranvier, so both answers would be accepted. Similarly with C, which points to the area of myelin within a Schwann cell — both answers are correct. There are several learning outcomes in this unit that deal with naming structures. Make sure that you can identify and label parts of neurones correctly. Making your own labelled drawings for your notes is one way to learn this.

Student A

(a) (ii) The resting potential of $-70\,mV$ is due to the distribution of ions across the axon membrane. It is maintained because Na^+ ions cannot diffuse through the membrane and there is a higher concentration of Na^+ outside the membrane, which makes the outside positively charged. There is a higher concentration of K^+ ions inside the neurone than outside, but inside there are negatively charged proteins and other compounds that cannot pass through the membrane. The sodium pump maintains the unequal distribution of Na^+ and K^+ by pumping out three Na^+ for every two K^+ pumped in.

Student B

(a) (ii) The sodium–potassium pump is responsible for the resting potential by pumping sodium ions out of the neurone and pumping potassium ions into the neurone.

ⓔ Student A has written a rather lengthy answer that could be written more concisely. The important points are:

- the unequal distribution of ions, with a higher concentration of Na^+ outside the membrane than inside and a higher concentration of K^+ inside than outside
- the sodium–potassium ion pump maintains this unequal distribution of ions
- the impermeability of the membrane to Na^+
- the voltage-gated channel proteins for Na^+ are closed

Student B has made one of these points so gains 1 mark.

Student A

(b) The results show the effect of a depolarisation at D on the transmission of impulses. In the first case (D1), the stimulus given by the electrode was below threshold, so no impulses were sent. In D2, the stimulus was above threshold and four impulses were recorded at E. A greater stimulus in D3 resulted in an increased frequency of impulses recorded at E. The action potentials recorded are all the same size. The strength of the stimulus is registered by the number of impulses sent per unit time (the frequency). This is known as the all-or-nothing rule.

Student B

(b) The results in the figure show the all-or-nothing rule. When they are stimulated, neurones either send impulses or they don't. In the first result, the stimulus is not big enough and no impulses are sent. In the second and third cases the membrane potential is bigger and impulses are sent by the motor neurone.

@ Student A has described and explained each result shown in Figure 3. The three separate results are identified using information in the figure (D1, D2 and D3), which saves time and space when writing the answer. Student A has also identified that this question is asking about threshold stimulation, the all-or-nothing rule and encoding information about the strength of stimuli by changing the frequency of impulses. The candidate has also used the data to give the relative frequency recorded at E in D3. Student B clearly understands the all-or-nothing rule, but has not referred to the difference in frequency between D2 and D3. Student B has not used all the data shown in the figure, but does get 2 marks for identifying the all-or-nothing rule and using the data to support this.

Student A

(c) Transfer impulses from one neurone to another in the correct direction.
Inhibit impulses from transferring from one neurone to another.
Filter out impulses that are below the threshold for the postsynaptic neurone.

Student B

(c) They can stop a neurone from sending impulses — inhibitory synapses.
They release transmitter substances, like ACh.
They make sure impulses travel one way through the nervous system.
Formation of synapses between neurones may be used in memory.

@ The question asks for three roles. Some candidates write in extra answers as has student B. The examiners mark only the first three answers, so even though student B's final point is correct, since synapses are thought to be involved in memory, it does not gain any credit. The second answer given by student B is not a *role* of a synapse — it is part of the mechanism by which they all work (see Figure 12 on p. 25). ACh is an acceptable abbreviation for acetylcholine and the examiners would recognise it, but it is advisable to write the term in full on the first occasion and put the abbreviation in brackets. You may then use the abbreviation in the rest of your answer. However, if you are asked to name a neurotransmitter you must write out the full name — acetylcholine. Remember that a synapse where acetylcholine is released is a cholinergic synapse. You must recognise and use the term cholinergic as it is given in one of the learning outcomes. Student B gains 2 marks for the first and third points.

@ **Student B gains 8 marks out of 15 for Question 2.**

Question 3 **Hormones**

Adrenaline is released by adrenal glands and stimulates the release of glucose from liver cells. The flow chart in **Figure 4** shows the events that follow the arrival of an adrenaline molecule at the surface of a liver cell.

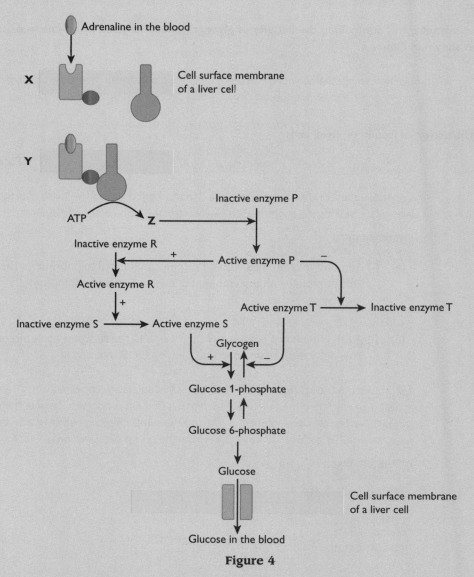

Figure 4

(a) Use the information in **Figure 4** to answer the following questions.

   (i) **Describe the events that occur when adrenaline arrives at the surface of a liver cell (X to Y).**   (2 marks)

  (ii) **Name compound Z.**   (1 mark)

  (iii) **Explain the meanings of the terms first messenger and second messenger.**   (2 marks)

ⓔ This is another question that asks you to use information in a figure. This means you should describe what you see in the diagram, look for compound Z and use what is shown to help explain the two terms.

(b) Enzyme S in Figure 4 is glycogen phosphorylase, which breaks down glycogen into glucose 1-phosphate.

Explain the advantage of controlling the activity of glycogen phosphorylase by an enzyme cascade, as shown in Figure 4. (3 marks)

ⓔ Remember that we need extra glucose in the bloodstream in a hurry. This might be the clue you need to answer this question. Always think around a question before you start writing.

(c) Describe the action of insulin on liver cells. (3 marks)

Total: 11 marks

ⓔ This is a test of knowledge. There are 3 marks, so you should give three effects of insulin on liver cells. Lowering the concentration of glucose in the blood is not one of them.

Student A

(a) (i) Adrenaline combines with its receptor in the cell membrane of the liver cell. This stimulates an enzyme that converts ATP into compound Z.

Student B

(a) (i) Adrenaline has a complementary shape to the receptor so combines with it. This activates the enzyme that converts ATP into cAMP.

ⓔ Both candidates gain 2 marks for identifying that adrenaline combines with a receptor and activates an enzyme in the cell surface membrane. Student A refers to Z rather than naming it, but that is acceptable here. Student B refers to the complementary nature of adrenaline and its receptor, which is good use of information from the cell signalling section of Unit F211.

Student A

(a) (ii) cyclic AMP

Student B

(a) (ii) cAMP

ⓔ Both answers are correct. You may use an accepted abbreviation here, such as cAMP. If you decide to use an abbreviation of your own in answers, then make sure you write out the full name first and put your abbreviation in brackets. You may want to do this when answering questions on the nerve impulse and refer to voltage-gated ion channels as VGIC.

OCR A2 Biology

Student A

(a) (iii) Adrenaline is the first messenger. It is secreted by endocrine glands (the adrenal glands) and travels in the blood to target cells in the liver. Adrenaline does not cross membranes — it combines with a receptor that stimulates a second messenger to act within cells and activate or inhibit cell processes. Cyclic AMP is the second messenger.

Student B

(a) (iii) Adrenaline is the first messenger and cAMP is the second messenger.

ⓔ Student A gives a full answer, making use of the examples of first and second messengers given in the figure. Student B gives examples of these terms, but does not explain their meanings. The candidate has not answered the question and fails to score.

Student A

(b) The advantage is that a large response can occur in a short period of time. Glucose is produced quickly and the glucose concentration of the blood increases quickly, which provides energy for muscle action — for example, when an animal has to escape from a predator. Enzymes P and R in the cascade each activate other enzymes so that lots of molecules of enzyme S are activated. This is amplification of the original stimulus from the first messenger, adrenaline.

Student B

(b) cAMP activates enzyme P which activates lots of molecules of enzyme R. When enzyme R is activated it activates lots of molecules of enzyme S. This is because enzymes can catalyse many reactions in a short time.

ⓔ Student A has given a full answer and, using the figure, has taken it as far as the release of glucose into the blood, which is the response to stimulation of the liver cell by adrenaline. Student B has spent time *describing* the enzyme cascade, rather than *explaining the advantage*. This is a common error: providing a standard definition, description or explanation rather than using this information to answer the question. Student A explains that the enzyme cascade amplifies the message delivered by adrenaline. This is why concentrations of hormones, such as adrenaline, insulin and glucagon, in the blood are so low — for example, the concentration of insulin in the blood is about $0.25\,\mu g\,dm^{-3}$.

In the cascade in Figure 4, the enzymes are as follows:

- P — protein kinase
- R — phosphorylase kinase
- S — glycogen phosphorylase
- T — glycogen synthase (or synthetase)

Enzymes catalyse many reactions in a short period of time. The number of reactions per unit of time is the turnover number. This means that a small number of molecules of adrenaline can activate a larger number of molecules of cAMP which activate protein kinase molecules and these in turn activate many phosphorylase kinase molecules. This is what happens during the cascade

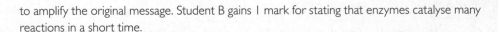

to amplify the original message. Student B gains 1 mark for stating that enzymes catalyse many reactions in a short time.

Student A

(c) Insulin stimulates liver cells to convert glucose to glycogen (glycogenesis) by activating the enzyme glycogen synthase (T on Figure 4). It also increases the uptake of glucose from the blood and its use in respiration. It stimulates the conversion of glucose into fat in the cells of adipose tissue.

Student B

(c) Insulin is a hormone that is secreted by the islets of Langerhans in the pancreas. It attaches to receptors on liver cell membranes and activates the conversion of glucose to glucogen so that it can be stored, and this lowers the blood sugar level. When the blood sugar level falls below the set point, glycagon is secreted by the pancreas to have the opposite effect.

ℯ Care has to be taken with this question since it asks for the effects of insulin on the *liver cells*, not just in general terms of decreasing the blood glucose concentration. This is one reason why student B has not done well — the answer contains information that is unnecessary. Student B has also made two spelling mistakes: 'glucogen' for glycogen and 'glycagon' for glucagon. Examiners may award marks when technical terms are misspelt, providing that the meaning is clear. However, there are some words, including glycogen and glucagon, that must always be spelt correctly because they can be confused. There was no need for student B to discuss glucagon in this answer, as it is irrelevant. The first sentence in the answer is also irrelevant. Student B fails to score. 1 mark would have been awarded for stating that insulin stimulates the storage of glycogen, but the misspelling costs the candidate this mark.

ℯ **Student B gains 4 marks out of 11 for Question 3.**

Question 4 Excretion

Table I shows the concentration of various components in the blood plasma, glomerular filtrate and urine of a human. The units are $g\,100\,cm^{-3}$.

Table 1

Substance	Plasma	Glomerular filtrate	Urine	Increase
Water	90.00	97.00	95.00	–
Protein	8.0	0.0	0.00	–
Glucose	0.1	0.1	0.00	–
Urea	0.03	0.03	2.00	
Uric acid	0.004	0.004	0.05	12-fold
Ammonia	0.0001	0.0001	0.04	400-fold
Na$^+$	0.32	0.32	0.35	No increase
K$^+$	0.02	0.02	0.15	7-fold

(a) Complete the table by calculating the factor by which the urea concentration has increased in the urine compared with its concentration in the blood plasma.

(I mark)

ⓔ If you are not sure how to do this, try one of the other calculations and see if you get the same answer as in the table.

(b) Explain why the concentration of ammonia in the blood plasma is very low.

(2 marks)

ⓔ Think about what the source of ammonia might be, what happens to it and why it is not a good idea to have a higher concentration in the blood. There are two ways to answer the question 'why'.

(c) Explain the figures for glucose and protein given in Table I.

(2 marks)

ⓔ Here you must refer to the numbers in each column even if they are zero. In fact, especially if they are zero because there is an important principle involved.

(d) The kidneys are involved in osmoregulation.

Explain how the kidneys are involved in controlling the volume of urine excreted.

(4 marks)

Total: 9 marks

ⓔ The term 'osmoregulation' is a clue here. Think about the hormone secreted to achieve conservation of water by the kidney. The volume of urine changes in response to this hormone.

Student A

(a) 67

Student B

(a) 66.67

e Both candidates have the correct result. When answering a question like this always look to see if there are similar calculations in the table so you can be sure you are following the same method. The examiner has not asked for the working of the calculation but it may be a good idea to show it:

$$\frac{2}{0.03} = 66.66$$

Student A has rounded up to the nearest whole number because other results are given like this in the table. The examiners may ask you to express your answer to the nearest whole number or to a certain number of decimal places. If not, express your answer to match other figures given in the table. Often examiners allow results to be given to a reasonable number of decimal places. They would not be willing to award a mark for 66.66666666, but would accept 66.$\dot{6}$ or 66.6 recurring.

Student A

(b) Ammonia is toxic and is converted into urea in the ornithine cycle in liver cells.

Student B

(b) Ammonia is poisonous.

e Both candidates gain a mark for stating that ammonia is poisonous or toxic. Student B has not noticed that there are 2 marks for this part-question; converting ammonia to urea is a suitable answer. Student A has realised that there are two quite different answers —that ammonia is harmful and that it is converted to something else.

Student A

(c) Glucose is small enough to be filtered but is reabsorbed in the proximal convoluted tubule. All of the glucose in the filtrate is reabsorbed. Protein molecules in the plasma are too big to be filtered so they remain in the blood, as is clear from the table.

Student B

(c) Glucose and protein are filtered but both are reabsorbed.

e Student B states that both substances are filtered. This is not the case for protein, so only 1 mark can be gained for giving the correct answer for glucose. In fact, some small protein molecules are filtered and are reabsorbed in the proximal convoluted tubule by endocytosis. However, the candidate is implying that *all the proteins* in the plasma are filtered and this is certainly *not* the case. Examiners would not expect you to know about these small proteins. They would expect you to know that the large plasma proteins are too big to be filtered and remain in the blood plasma.

Student A

(d) The volume of urine produced is controlled by the collecting ducts (CDs) in the kidney. When the body has plenty of water and the water potential of the blood is slightly higher than it should be, the walls of the CDs remain impermeable to water. Therefore, the dilute urine that comes from the distal convoluted tubules flows straight through to the pelvis, the ureter and the bladder. When dehydrated, the water potential of the blood is lower than the norm and ADH is secreted. This stimulates vesicles with aquaporins (water channels) to fuse with the membranes in the CD cells. Water passes by osmosis out of the dilute urine into the blood and so the concentration of the urine increases and, because water has been reabsorbed, the volume of urine is less than it would have been. Water is conserved.

Student B

(d) This depends on how much water is drunk. If a lot of water is drunk then there will be lots of water in the blood that needs to be excreted. This means that more blood will be filtered in the kidneys and the filtrate will pass through the PCT and into the collecting ducts, which do not absorb any water. When there isn't much water in the blood then less blood is filtered and more water is reabsorbed in the collecting ducts.

ⓔ Student B makes a common error about filtration and scores no marks. The rate of filtration in the kidneys is fairly constant — it does not change depending on the water potential of the blood. The rate of filtration in the glomeruli of both kidneys is about $125\,cm^3\,min^{-1}$. The volume and concentration of the urine are controlled by the collecting ducts, as explained by student A. The volume of urine collected from an individual depends on how often the bladder is emptied, so it is important to know how long a urine sample has been collecting in the bladder. The concentration of a sample of urine depends on the conditions during the time it has been collecting. It is possible that initially much ADH was secreted and concentrated urine was produced, but later less was secreted so the concentration of urine decreased. The concentration of the sample would be an average of the concentrations produced over this time.

ⓔ **Student B gains 3 marks out of 9 for Question 4.**

Question 5 **Photosynthesis**

Figure 5 is an outline of the cycle of reactions that occurs in the light-independent stage of photosynthesis.

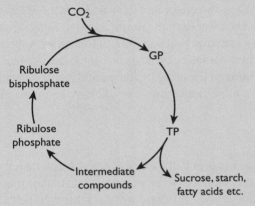

Figure 5

(a) (i)   State the precise site in the chloroplast of the light-independent stage of photosynthesis.   (1 mark)

> **e** Think about a diagram of the chloroplast before you commit yourself to an answer to this question. This is one reason why it is a good idea to make detailed annotated diagrams on large sheets of paper to help your revision.

(ii) Show on Figure 5 where reduced NADP and ATP, the two products of photophosphorylation in the light-dependent stage of photosynthesis, are used in the light-independent stage.   (1 mark)

> **e** This type of question does not have any dotted lines on the examination paper for your answer, so it is easy to miss. Always check through your paper and make sure you have an answer for all the marks shown in brackets on the right-hand side of the page.

(iii) Explain why the concentration of GP in chloroplasts increases as light intensity decreases at sunset.   (3 marks)

> **e** The diagram shows the Calvin cycle of the light-independent stage. Light intensity relates to the light-dependent stage. This is the clue to unlocking this question.

In 1961, the English scientist, Peter Mitchell, proposed the theory of chemiosmosis to explain how ATP is synthesised in chloroplasts and mitochondria. Figure 6 shows an experiment with grana isolated from chloroplasts that was carried out by André Jagendorf and co-workers and published in 1963.

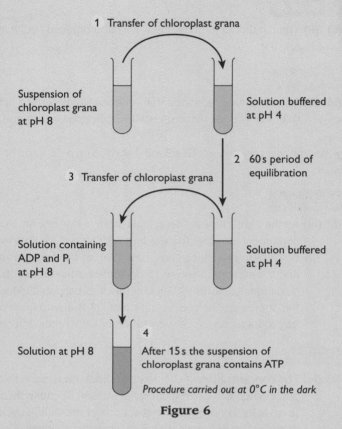

1 Transfer of chloroplast grana

Suspension of chloroplast grana at pH 8

Solution buffered at pH 4

2 60 s period of equilibration

3 Transfer of chloroplast grana

Solution containing ADP and P_i at pH 8

Solution buffered at pH 4

Solution at pH 8

4
After 15 s the suspension of chloroplast grana contains ATP

Procedure carried out at 0°C in the dark

Figure 6

(b) Explain how the experiment shown in Figure 6 provides evidence for the theory of chemiosmosis. If you wish, you may refer to the steps in the procedure by the numbers given in Figure 6.

(4 marks)

Total: 9 marks

ⓔ The key word here is chemiosmosis. Think about your knowledge of chemiosmosis — membrane, energy transfer, electron flow, proton pumping, proton gradient — and see how the experiment supports those ideas.

Student A

(a) (i) Stroma

Student B

(a) (i) Stoma

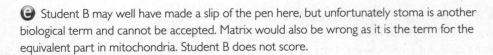

(e) Student B may well have made a slip of the pen here, but unfortunately stoma is another biological term and cannot be accepted. Matrix would also be wrong as it is the term for the equivalent part in mitochondria. Student B does not score.

Student A

(a) (ii) The candidate indicated that ATP and reduced NADP are used between GP and TP.

Student B

(a) (ii) The candidate indicated that reduced NADP is used between GP and TP and that ATP is used between ribulose phosphate (RP) and RuBP.

(e) Both answers are correct. Check with Figure 36 on p. 56 to confirm this. Student B gains 1 mark.

Student A

(a) (iii) As the light intensity decreases there is less energy to drive the light-dependent stage so there is less ATP produced. This means that less GP is converted into triose phosphate (TP), so the concentration of GP increases and the concentration of TP decreases. Also the reaction catalysed by rubisco (see Figure 5, RuBP → GP) continues as carbon dioxide is still available and this increases the concentration of GP. But with decreasing TP, less RuBP can be produced so the concentration of GP will reach a constant level.

Student B

(a) (iii) The concentration of GP increases because there is less photosynthesis taking place when it is getting darker at sunset. Because there is less photosynthesis there is less energy for the reactions of the Calvin cycle which fix carbon dioxide. The enzymes therefore work more slowly.

(e) This question is asking about the changes in concentrations of GP, TP and RuBP with changing light intensities. The graph below shows these changes.

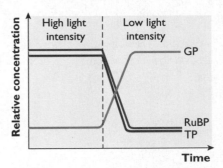

Student A's answer explains the effect of light intensity as shown in the graph. You may also be asked to explain the effect of changing the carbon dioxide concentration on the relative quantities of the same three compounds. Here is the graph.

OCR A2 Biology

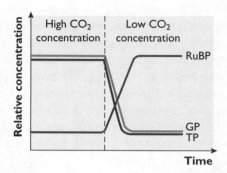

When the carbon dioxide concentration decreases there is less carbon dioxide to fix. Look at the outline diagram of the Calvin cycle in Figure 5. If there is less carbon dioxide, then RuBP accumulates because it is not being used to fix carbon dioxide. If carbon dioxide is not being fixed, then GP will not be formed, so the concentration decreases as will that of TP. In the first graph above, the explanation is to do with the provision of energy from the light-dependent stage. In the second graph, it is to do with carbon dioxide fixation in the light-independent stage. If you remember the outline of the Calvin cycle (see Figure 5) and the three processes that take place (see p. 55), then you have the means to explain these two graphs.

In answering the question, student B states that less photosynthesis takes place when it gets darker and then states that there is less energy available. This is too general an answer. Examiners expect candidates to use information about ATP and reduced NADP from the light-dependent stage, especially as there is a hint about this in (a)(ii). The examiners want to know about the reactions of the Calvin cycle. Student B's answer is not precise enough to gain any credit.

Student A

(b) Chloroplasts have been broken apart to release the grana which are in suspension. Each granum is made up of stacks of thylakoids. When the thylakoids are put into the solution at pH 4, protons move down their concentration gradient into the thylakoid spaces. This gives the inside of the thylakoid spaces a low pH (as they would normally have due to proton pumping, using energy from the transport of electrons from the photosystems). The grana are transferred into a solution of pH 8, which has a low concentration of protons and is like the stroma in the intact chloroplasts. There is now a proton gradient from the thylakoid space to the stroma. ATP is formed at stage 4 by protons diffusing through ATP synthase to form ATP from the ADP and phosphate added at stage 3.

Student B

(b) Grana are the site of the light-dependent stage of photosynthesis. When light strikes photosystem 2, electrons travel along the electron transport chain to photosystem 1. The energy released during the transfer of electrons is used to pump protons from the stroma into the thylakoid space. Protons can only diffuse through ATP synthase. This means that the thylakoid space has a low pH and protons diffuse through ATP synthase into the stroma down their concentration gradient and ATP is formed on the stromal side of the thylakoids ready for the light-independent stage.

(e) This is the so-called 'acid bath' phosphorylation procedure that was carried out in the early 1960s and provided evidence for the chemiosmotic theory proposed by Mitchell. This is a good example of the type of question that addresses the *How Science Works* aspect of the specification. You can expect to be asked to explain how the results of an experimental procedure support or refute a statement, theory, hypothesis or prediction (as in Question 1(a)(ii)). Note that the whole procedure was carried out in the dark, so there was no involvement of the photosystems and the proton pump. Student A has given a thorough answer, but student B has described chemiosmosis without answering the question. Student B scores 2 marks for the use of appropriate information about chemiosmosis: the correct orientation of the proton gradient and the production of ATP by ATP synthase.

There are various aspects to the *How Science Works* theme in the specification, some of which concern the nature of scientific evidence. It is always worth asking how we know about the topics in the specification and how the research was carried out. This is particularly important for learning outcomes identified as matching the different aspects of *How Science Works*. If you look carefully at the specification, you will see that they have been identified for you.

(e) **Student B gains 3 marks out of 9 for Question 5.**

Question 6 **Respiration**

Figure 7 shows a metabolic pathway that occurs in mammalian muscle tissue.

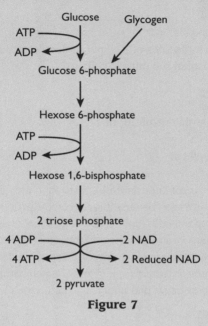

Figure 7

(a) (i) Name the metabolic pathway shown in Figure 7.   (1 mark)

🅔 This shows how important it is to make a large diagram of respiration and annotate it.

(ii) State the net yield of ATP when one molecule of glucose is metabolised by this pathway.   (1 mark)

🅔 Notice that this question asks for the *net* yield.

(iii) State where in muscle tissue this pathway occurs.   (1 mark)

🅔 Try to be as precise as possible when answering questions that ask 'where'.

The concentration of NAD in muscle tissue is very low, about 0.8 µmol g^{-1} of muscle tissue. In aerobic respiration, reduced NAD is converted to NAD by mitochondria.

(b) (i) With reference to Figure 7, explain how reduced NAD is recycled in muscle tissue when oxygen is not available.   (3 marks)

🅔 You could answer this by continuing the pathway to show how reduced NAD is recycled.
Write this in the answer space. A diagram on its own will not get the marks, but it will help.

(ii) Explain how reduced NAD is recycled when oxygen is available in the muscle tissue.   (2 marks)

Total: 8 marks

ⓔ The clue in the question is 'oxygen'. Respiration with oxygen is aerobic. This should help you first to locate the part of respiration where reduced NAD is recycled. Then, you can explain briefly how this happens.

Student A

(a) (i) Glycolysis
(ii) Two molecules of ATP per molecule of glucose
(iii) Cytosol in the muscle cells

Student B

(a) (i) Anaerobic respiration
(ii) Four
(iii) Cytoplasm

ⓔ Student B has confused anaerobic respiration with glycolysis. The pathway shown cannot be anaerobic respiration because it is not complete. Glycolysis is the metabolic pathway that is common to *both* anaerobic *and* aerobic respiration. Anaerobic respiration in muscle tissue has another reaction, not shown in Figure 7, but described in student A's answer to part (b). Cytoplasm is the site of glycolysis given in the specification, so student B gains 1 mark for part (a) (iii). The more precise answer to part (iii) is cytosol. Remember that mitochondria, where the link reaction, Krebs cycle and oxidative phosphorylation occur, are also part of the cytoplasm.

Student A

(b) (i) Pyruvate acts as the hydrogen acceptor as it receives hydrogen from reduced NAD. This means that reduced NAD is oxidised and is now available for the reaction that occurs in glycolysis and is shown in the pathway. NAD is a coenzyme. When pyruvate is reduced it is converted to lactate which diffuses out of the muscle tissue into the blood:

Pyruvate + reduced NAD → lactate + NAD

Student B

(b) (i) There is no oxygen available, so pyruvate does not enter the mitochondria to be respired aerobically. Instead, it is converted to lactate which leaves the muscle.

ⓔ Student A gives a full answer. It is always a good idea to give an equation or draw part of a metabolic pathway if it helps your answer. However, do not *describe* a pathway if you are asked to explain it. Student B gains 1 mark for stating that pyruvate is converted to lactate.

Student A

(b) (ii) When oxygen is available, reduced NAD is recycled by the mitochondria. It is oxidised by intermediate compounds which pass into the matrix of the mitochondrion. The hydrogens from reduced NAD from glycolysis are passed to the ETC.

(b) (ii) Oxygen is the final electron acceptor in aerobic respiration. This is how reduced NAD is recycled.

ⓔ Student B has jumped to the end of the story! Student A makes it clear that reduced NAD is recycled by the action of mitochondria, thus gaining the first of the 3 marks available. It is not necessary to know the details of the recycling of NAD from glycolysis (such as the names of the intermediate compounds), but you should know that the hydrogen ions and electrons from reduced NAD are made available to the electron transport chain (ETC). Student A uses an accepted abbreviation here.

ⓔ **Student B gains 2 marks out of 8 for Question 6.**

Overall, Student B gains 23 marks. This is not enough for an E grade on the paper as a whole.

You can see that Student B has lost marks for a number of different reasons:

• The mark allocation has not been followed, e.g. Q4(b), Q6(b)(ii).
• Technical terms are spelt incorrectly, e.g. Q3(c) and Q5(a)(i).
• More responses than required have been given, e.g. Q2(c).
• Some answers are not developed fully, e.g. Q2(a)(ii) and (b).
• Terms from the specification are used incorrectly, e.g. Q1(b), Q6(a)(i).
• Terms have not been explained, e.g. Q3(a)(iii).
• Information has been described, rather than explained, e.g. Q3(b), Q5(b).
• Not understanding what is required for an answer, e.g. Q1(a)(ii), Q4(d).
• Instructions have not been followed carefully, e.g. Q1(a)(ii) where examples have been given instead of an explanation.
• Data provided have not been used fully, e.g. Q2(b).
• Answers are not precise enough, e.g. Q5(a)(iii) where the candidate uses the word photosynthesis rather than referring to the energy provided by the light-dependent stage.
• The 'full story' has not been given, e.g. Q6(b)(ii).
• Answers are too general and need more development, e.g. Q6(b)(i).
• In questions that address How Science Works, not answering the question to explain how data support or refute statements, e.g. Q1(a)(ii) and Q5(b).
• Errors have been made because a process has not been understood, e.g. Q4(c) and (d).
• Irrelevant material is included, e.g. Q3(c).

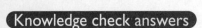

Knowledge check answers

1 Ectotherms gain heat from their surroundings; endotherms generate heat in their bodies, e.g. by respiration and shivering.

2 Homeostasis is the maintenance of near constant internal conditions in the body.

3 Hypothalamus

4 It reduces heat loss by physiological and behavioural means. A mammal conserves heat by contracting hair-erector muscles to give a greater depth of fur, reducing blood flow through the outer capillaries of the skin and curling up to reduce the surface area exposed to the cold air. It also generates heat by shivering and increasing the rate of metabolism in the liver. Some mammals have brown fat which generates heat.

5 Negative feedback acts to maintain a factor at a near constant level; positive feedback acts to continually increase the factor. Negative feedback occurs continually; positive feedback is a short-lived mechanism.

6 Receptors convert the form of energy of a specific stimulus into electrical impulses in neurones.

7 Three factors influence the speed of conduction in neurones. Conduction is much faster in myelinated neurones than in unmyelinated neurones. Conduction along myelinated neurones is faster because the impulse jumps from node to node. Conduction is faster in wide neurones than in narrow neurones because resistance decreases as the cross-sectional area increases (you might know this from your knowledge of physics). Speed of conduction is also influenced by body temperature; it is faster in animals with high body temperature as the movement of ions through channel proteins and their opening and closing is temperature-dependent.

8 Sensory neurones transmit impulses from receptors to the central nervous system (CNS); relay neurones transmit impulses between sensory and motor neurones; motor neurones transmit impulses from the CNS to effectors (muscles and glands). Avoid using the terms 'message' and 'signal' when writing about neurones.

9 110mV. This varies between neurones depending on the resting potential and the maximum potential difference at the height of the action potential.

10 There are several answers to this:
   - Impulses go from receptor to CNS or from CNS to effector; there is no point in information going in the reverse directions because receptors receive stimuli and effectors make the changes required.
   - Synapses are polarised — there are vesicles of neurotransmitter on one neurone and receptors for these cell-to-cell signalling compounds on the other. Therefore, nerve impulses go in one direction.
   - The region of a neurone behind an action potential is undergoing the refractory period; voltage-gated sodium ion channel proteins are inactivated so cannot open. Therefore, the impulse cannot go backwards.

11

	Voltage-gated ion channel proteins	
Stage of action potential	Sodium	Potassium
During depolarisation towards threshold	Activation gates on some channel proteins opening	All activation gates closed
Rising phase	More activation gates opening	Most activation gates closed, some opening slowly
Falling phase	Inactivation gates closing quickly	Activation gates opening slowly
Refractory period	All inactivation gates closed	All activation gates open

12 Depolarisation stimulates voltage-gated calcium ion channel proteins to open allowing calcium ions to flow into the synaptic bulb. This stimulates vesicles to move towards the presynaptic membrane. These vesicles fuse with the membrane releasing neurotransmitter molecules into the synaptic cleft (exocytosis). Neurotransmitter molecules diffuse across the synapse and bind with ligand-gated sodium ion channel proteins; sodium ions diffuse into the postsynaptic membrane.

13 Exocrine — secretion into a duct; substance(s) secreted travel down a duct to the outside (sweat duct) or into another organ (liver and pancreas into duodenum). Endocrine — secretion into the blood (not into a duct).

14 Insulin stimulates the uptake of glucose from the blood and its storage as glycogen; glucagon stimulates the breakdown of glycogen to produce glucose that diffuses into the blood.

15 The table below has some points of comparison. You may be able to think of others.

Feature	Glycogen	Glucagon
Type of biochemical	Carbohydrate Polysaccharide	Protein Single polypeptide
Molecular structure	Branched	Unbranched
Monomer	α-glucose	Amino acids
Function	Energy storage	Signalling molecule (hormone)
Site of production	Liver cells, muscle cells	α-cells in the islets of Langerhans in the pancreas

16 Type 1 diabetes is caused by an inability to produce insulin; type 2 diabetes is the inability of target cells to respond to insulin.

17 Stem cells could be placed into the body where they would differentiate into islet β-cells and secrete insulin. To protect against destruction by the immune system they could be encapsulated before transplant.

18 18.9/19%. The volume of plasma that passes through the glomeruli every minute is $660 \, cm^3$.

19 Glucose, ions (e.g. sodium and chloride), urea and water are reabsorbed, so the volume of filtrate decreases significantly. Reabsorption occurs by active transport (glucose and sodium ions), by diffusion (urea) and by osmosis (water).

20 The pore in the centre of the aquaporin is just large enough for water but does not let larger molecules or ions through. The channel is lined by amino acid residues that are charged and repel ions of the same charge.

21 The relative molecular mass of insulin is below 69 000 so it is filtered through the glomerulus and is not reabsorbed.

22 Calculate the cross-sectional area of the capillary tube (πr^2) and multiply this by the distance travelled by the air bubble.

23 At low light intensities and low carbon dioxide concentrations, there is a very low rate of photosynthesis. The rate of respiration is faster than the rate of photosynthesis. The oxygen produced is used by mitochondria in the plant cells so no oxygen is released. As the light intensity and concentration of carbon dioxide increase the rate of photosynthesis increases, but the rate of respiration stays constant. After the point when the rate of photosynthesis becomes faster than the rate of respiration, there is excess oxygen produced which diffuses out of the plant. The rate of respiration remains constant because the temperature is constant (it is a control variable).

24 Extra lighting to increase light intensity; heaters to increase temperature; ventilation to decrease temperature (too hot and enzymes denature); carbon dioxide enrichment by burning gas or releasing carbon dioxide (by-product of fermentation industries).

25 Respiration is the oxidation of complex carbon compounds with the transfer of energy to ATP. Gaseous exchange is the diffusion of carbon dioxide and oxygen across the gas exchange surface in the lungs between air and the blood.

26 ATP is involved in energy transfers within the cells of all living organisms. It is formed in respiration and photosynthesis and used in energy-consuming processes, such as active transport and biosynthesis.

27 Cytosol, inner mitochondrial membrane, thylakoid membrane

28 Malonate is a competitive inhibitor of the enzyme. You studied enzyme inhibitors in F212. You should expect questions on some of the content of F211 and F212 in your F214 paper.

29 Glycolysis — cytosol; link reaction and Krebs cycle — matrix of mitochondrion; oxidative phosphorylation — inner mitochondrial membrane

30 NAD and FAD are hydrogen carriers; coenzyme A transfers two-carbon acetyl groups into the Krebs cycle.

31 Lactate is an energy-rich molecule; excreting it would be a waste of energy; cardiac muscle respires lactate; it can be converted into glycogen and stored for future use.

32 Lipids are energy-rich because they are highly reduced with a much larger proportion of hydrogen atoms to carbon than carbohydrates. On oxidation during respiration they release more energy as more reduced hydrogen carriers are produced per mole. This is another synoptic question from F212.

W

Y

Z